MENTAL HEALTH: PHILOSOPHICAL PERSPECTIVES

PHILOSOPHY AND MEDICINE

Editors:

H. TRISTRAM ENGELHARDT, JR.

Georgetown University, Kennedy Institute, Washington, D.C., U.S.A.

STUART F. SPICKER

University of Connecticut Health Center, Farmington, Conn., U.S.A.

VOLUME 4

MENTAL HEALTH: PHILOSOPHICAL PERSPECTIVES

PROCEEDINGS OF THE FOURTH TRANS-DISCIPLINARY SYMPOSIUM ON PHILOSOPHY AND MEDICINE HELD AT GALVESTON, TEXAS, MAY 16–18, 1976

Edited by

H. TRISTRAM ENGELHARDT, JR.
Georgetown University, Kennedy Institute, Washington, D.C., U.S.A.

and

STUART F. SPICKER
University of Connecticut Health Center, Farmington, Conn., U.S.A.

SPRINGER-SCIENCE+BUSINESS MEDIA, B.V.

Library of Congress Cataloging in Publication Data

Trans-disciplinary Symposium on Philosophy and Medicine,
4th, Galveston, 1976.
Mental health, philosophical perspectives.

(Philosophy and medicine; v. 4)
Includes bibliographies.
1. Mental health – Con^^resses. 2. Psychology
Pathological – Congresses. I. Engelhardt, Hugo Tristram,
1941– II. Spicker, Stuart F., 1937– III.
Title. [DNLM: 1. Mental health – Congresses. 2. Mental
disorders – Congresses. 3. Philosophy, Medical – Congresses.
W/PH609 v. 4/WM100 T772m 1976]
T772m 1976]
RA790. A1T72 1976 616.8'9'001 77–24974
ISBN 978-94-015-6911-8 ISBN 978-94-015-6909-5 (eBook)
DOI 10.1007/978-94-015-6909-5

TABLE OF CONTENTS

INTRODUCTION

The concept 'health' is ambiguous [18, 9, 11]. The concept 'mental health' is even more so. 'Health' compasses senses of well-being, wholeness, and soundness that mean more than the simple freedom from illness – a fact appreciated in the World Health Organization's definition of health as more than the absence of disease or infirmity [7]. The wide range of viewpoints of the contributors to this volume attests to the scope of issues placed under the rubric 'mental health.' These papers, presented at the Fourth Symposium on Philosophy and Medicine, were written and discussed within a broad context of interests concerning mental health. Moreover, in their diversity these papers point to the many descriptive, evaluative, and, in fact, performative functions of statements concerning mental health. Before introducing the substance of these papers in any detail, I want to indicate the profound commerce between philosophical and psychological ideas in theories of mental health and disease. This will be done in part by a consideration of some conceptual developments in the history of psychiatry, as well as through an analysis of some of the functions of the notions of mental illness and health.

'Mental health' lays a special stress on the wholeness of human intuition, emotion, thought, and action. This positive sense of mental health as soundness of perception and conduct, not just freedom from mental disease (the negative or privative sense of mental health), endows 'mental health' with its particular rhetorical flourish and force. Being mentally ill bears on the integrity of the person afflicted in a way in which being physically ill does not. One can be a diabetic or have heart disease and maintain a distance from the pathological processes of one's body in a way one cannot if one is schizophrenic or has an obsession. This is not to say that physical illnesses do not intrude into one's psyche or fashion one's view of the world. The diabetic who is hypoglycemic or acidotic may experience alterations in clarity of consciousness; the individual with angina finds his or her life structured by limitations on activity imposed by a pain that is insistent and intrusive. But the schizophrenic experiences a disintegration of his or her very self, a coming apart and loss of control, a feeling of the world as alien. The person with an obsessive compulsive neurosis is impelled to perform activities or finds his or her mind intruded upon by unwanted ideas.

H. T. Engelhardt, Jr. and S. F. Spicker (eds.), Mental Health: Philosophical Perspectives, VII–XXII.

As a result, mental disease or illness strikes to the core of persons in a more direct fashion than most physical illnesses. This is especially true of psychoses, and, in particular, thought disorders. Schizophrenia can be seen as a failure to integrate reality, to construct a coherent framework of experience. The conceptual importance of the integration of reality is salient in the history of the development of contemporary psychiatry and psychology. Eugen Bleuler in his classic work on schizophrenia, *Dementia Praecox*, while sketching the hallmarks of the disease – ambivalence, inappropriate affect, associative disorder, and autims – indicated, especially through the last criterion, an inability to gear into the common reality of everyday life. 'Healthy people have a tendency, in logical operations, to draw upon all appropriate material without consideration of its affective value. On the other hand, the schizophrenic loosening of logical processes leads to the exclusion of all associations conflicting with mentally charged complexes' ([2], p. 373). Instead of attempting to integrate all the data of experience, the schizophrenic attends to some and ignores others in order to meet idiosyncratic emotional needs. The result is a picture of reality not shared, or not fully shared, in common with other persons.

In developing his view of mental health and illness, Bleuler signalled the extent to which mental health indicates a successfully intersubjective construction of reality. In his *Textbook on Psychiatry*, Bleuler argued that it is impossible to prove the existence of an external world: '. . . for the existence of the external world there are no proofs. That the table which we see has existence is only an assumption, even if of practical necessity' ([3], p. 8). Mental health includes apparent acceptance of participation in an assumed objective reality. Mental health becomes synonymous with the ability to gear into a commonly constructed world of everyday life and experience. '[I]f I once take for granted the existence of the table, and that of other people, and the external world, then this table can be shown to these other people' ([3], p. 8).

For Bleuler, the failure to participate in this intersubjective reality shows itself in the autistic mentation of schizophrenics, a failure to gear into everyday life and an emersion in a world not structured by those logical constraints which make an intersubjective world possible. As Bleuler put it:

> In the same way as autistic feeling is detached from reality, autistic thinking obeys its own special laws. To be sure, autistic thinking makes use of the customary logical connections insofar as they are suitable, but it is in no way bound to such logical laws. Autistic thinking is directed by affective needs; the patient thinks in symbols, in analogies, in fragmentary concepts, in accidental connections. Should the same patient turn back to reality, he may be able to think sharply and logically ([2], p. 67).

Mental health thus includes the ability and the commitment to participate in an intersubjective reality bound by the general constraints of rationality.

Such views of schizophrenia, and *pari passu* of mental health, have been stressed by others. For example, Silvano Arieti summarized von Domarus' principle [23], in this fashion: '*Whereas the normal person accepts identity only upon the basis of identical subjects, the paleologician* (e.g., the schizophrenic) *accepts identity based upon identical predicates*' ([1], p. 194). Thus, 'suppose that the following information is given to a schizophrenic: "The President of the United States is a person who was born in the United States. John Doe is a person who was born in the United States." In certain circumstances, the schizophrenic may conclude: "John Doe is the President of the United States" ' ([1], p. 194). Or, ' "The Virgin Mary was a virgin; I am a virgin; therefore, I am the Virgin Mary" ' ([1], p. 195). Though this account of the thought disorder of the schizophrenic does not apply to the difficulty of neurotics, nor apply in the same way to other psychoses, or even all schizophrenics, it indicates the extent to which being in mental health, unlike being in physical health, turns immediately and explicitly on questions concerning the nature of reality and of proper attitudes towards that reality. As a consequence, mental health can be seen in theoretical terms as it is in common discourse – namely, indicating an adequate grasp upon reality.

But how 'adequate' does one's grasp upon reality need to be for mental health? And where does one draw the lines distinguishing mental health, 'normal' idiosyncrasies, and mental disease? These questions are addressed in a classic article by D. Hack Tuke, 'Imperative Ideas', where the author discussed the obsessions and compulsions of many individuals who were well integrated into their societies, including Dr. Samuel Johnson. Tuke concluded that individuals afflicted by 'imperative ideas,' by obsessions and compulsions, can hardly be termed persons 'of perfectly sound minds.' But such afflictions do not preclude legal competence. For example: 'no "last will and testament" would be set aside in this case (that of a law student obsessed concerning the correct placement of negatives in sentences), or on the ground that a testator had an invincible desire to touch certain objects (e.g., Dr. Samuel Johnson) . . . ' ([22], p. 191). Further, in the case of such difficulties, which do not incapacitate an individual's ability to deal with reality, how does one distinguish mental disease from personal peculiarities compatible with mental health? Consider Dr. Tuke's suggestion that an inclination to speculate inquiry into cosmogony is a form of insanity, or at least a form of the symptom 'insanity of doubt' or 'maladie du doute.'

Closely allied is the insanity of the metaphysicians – Schöpfungsfrage. According to the late Professor Ball, of Paris, the malady of the metaphysicians constitutes a subdivision of the malady of doubt ([22], p. 180).

Though Ludwig Wittgenstein indicated that such questions were unhealthy, and in need of philosophical therapy ('The philosopher's treatment of a question is like the treatment of an illness' [25]), he did not conflate such 'unhealth' and 'therapy' with 'mental disease' and 'psychiatric therapy.' As Thomas Szasz argues in this volume, the disposition to such conflation is one of the singular dangers in the use of such terms as 'mental health' and 'mental disease.' Since mental health turns on judgments concerning the wholesomeness of particular views or perspectives on reality, the term 'mental disease' can be used, among other things, to indict political heterodoxy and to maintain the control of a political establishment. In contrast with somatic diseases which are imposed upon us by nature, Professor Szasz argues that mental diseases are really problematic ways of living which are assumed by individuals [20]. Mental disease is a myth which allows the reclassification, for socially useful purposes, of difficulties in living for which individuals (often called 'patients') are responsible.

The view of mental health as a way of life which expresses a preference for 'common unhappiness' over 'hysterical misery,' to recall Freud's words [17], is found in the contemporary roots of psychiatry. In fact, Freud himself, who is often presented as a psychic determinist, advanced psychoanalysis as a means for making such choices between mental health and disease possible. '. . . analysis does not set out to make pathological reactions possible, but to give the patient's ego *freedom* to decide one way or the other' [12]. This picture of psychoanalysis accords well with that forwarded by Thomas Szasz when he contends that 'the purpose of psychoanalysis is to give patients constrained by their habitual patterns of action greater freedom in their personal conduct' [21].

'Mental health' comes to designate the ability to act freely and rationally. Autonomous moral agency is the ideal of mental health. Of course, much is included in such a characterization – and at the same time very little. On the one hand, it may be useful to be reminded that being a person requires rationality and freedom in action, and that processes in a person which hinder such are usually termed 'diseases.' Since health is in part a function of the adaptation of an organism to an environment, human health, and especially mental health, will be influenced by the surrounding culture – the human environment in its completeness. But beyond that, what will count as a disease (e.g., homosexuality) will turn, at least in part, on cultural norms. On the other hand, any and all rational free action is embedded in a particular culture with a particular history. Man is a symbol-using animal whose environment is a mesh of symbols, a culture, a point indicated in this volume not only by the contribution of Professor Szasz, but also by that of Professor Fabrega. The lineaments of rationality and freedom will be drawn in detail by the social forces within a particular culture.

Still, one central point is missing – the mind lives in and through the brain. Up to this point in this introduction, minds and their diseases have been discussed without reference to bodies or brains. Mental health has been sketched in terms of the states of minds alone. Yet, mental diseases not only allow of physical therapies, but also a great number of mental illnesses, including the organic brain syndromes, attend dysfunctions of the central nervous system. Many investigators are convinced from empirical, as well as philosophical considerations, that all mental diseases have a physical basis. In one sense, such a contention is harmless. Even if one is a Cartesian, one will admit that minds act through bodies, in fact, through brains. But if one takes seriously that being in the world is always physical, one must also accept that all mental activity has a physical reality [10]. Still, to say that all mental life has a physical reality is not to agree that all mental illness has somatic causes, unless one means something trivial such as that all symbols have some sort of physical bearer. It is one thing to say that early home environment causes schizophrenia, and another to hold that schizophrenia is due to an immunological defect (even if home environments are in some 'basic' sense physical).

We are brought, then, to distinguishing among somatic, psychological, and, in fact, social models of mental health. Somatic models of mental health would include those accounts of mental health that place an emphasis upon genetic and other physical factors. A recent example of such is the account of schizophrenia proffered by Robert G. Heath, that the condition develops because of a genetically based error of metabolism [15]. While somatic accounts have a history in medicine that extends to the humoral theory of disease, psychological models are perhaps currently the most prevalent. The impetus for giving accounts of mental health in terms of psychodynamics arose in large part through the impact of the work of Sigmund Freud. Freud's theories, it should be noted, developed at the end of an era impressed by the somatic basis of insanity. Social models are found in the work of persons such as Harry Stack Sullivan, who in his *Interpersonal Theory of Psychiatry* argued for the study of 'the interpersonal situations through which persons manifest mental health or mental disorder' [9]. In social models, disease and health are seen as societal phenomena, not primarily as states of individual humans. Within such models, it is social interactions that are healthy or diseased, rather than particular persons considered in isolation.

Each of these models captures only a dimension of the reality of mental health and disease, in that all of mental life has a somatic presence and a social context [17]. Consider the somatic dimension. To be in the world is always to be somewhere, and being somewhere is always physical. Moreover, much that we know attests to the dependence of the mind on the brain – e.g., drinking alcohol, taking a tranquilizer, having a cup of coffee, etc. One must acknowledge mental life in health and illness as always possessing a physical substrate. Still, some accounts of mental disease are more somatic than others in advancing causes of mental disease which have no mental aspect (e.g., inborn errors of metabolism). Other accounts, such as that of Freud's, involve only a claim that there must be patterns of neurophysiological dysfunction that accompany the psychological dysfunctions of mental illness [13]. A psychological account can be given in independence of a physiological account of mental illness – as also exemplified by Bleuler. Finally, social theories of mental health and disease give accounts of mental phenomena in terms of social realities and forces, and can be viewed as especially full or complete portrayals of mental health and disease. They place physical and psychological processes in their larger, more ample framework – the network of social interactions.

To talk of physical, psychological, or social causes of mental health and disease is thus to identify elements of particular accounts which tell the best or the most complete account about how such states of affairs come about. As a consequence, to say that someone is suffering from a depressive neurosis due to the loss of a love object is to say more than that a set of stimuli has evoked an identifiable response pattern, even if, in a sense, all stimuli and responses have physical bases. The story of the depression is not complete until the psychological side has been told, even if the depression is curable through non-psychological modalities of treatment. Such a depression (i.e., one 'caused' by the loss of a love-object) is psychologically caused in a way that Korsakov's psychosis is not – there is not the same need for a psychological story in order to account for Korsakov's psychosis as is the case with respect to exogenous depressions. Explaining a particular individual's Korsakov's psychosis may require examining psychological circumstances – but such circumstances are not dimensions of the precipitating causes of an exogenous depression. Still, a more ample picture of the etiology of the Korsakov's psychosis would include an account of the psychological and sociological factors leading to alcoholism.

One should also note that accounts of physical or somatic health and disease can demonstrate the same sorts of complexities. Consider Meyer

Friedman and Ray H. Rosenman's classic account of the relationship between being a hard-driving, 'Type A' person, and being subject to a high risk of coronary artery disease [14]. In fact, there appears to be evidence that psychological stress is linked with many physical diseases [16] – e.g., loss of a spouse or a divorce is associated with the development of cancer [25]. Which is to say, somatic and mental diseases and health cannot be clearly distinguished on the basis of the causes involved. Both somatic and psychological variables appear to correlate with the development of somatic and psychological diseases. Physical diseases have mental components (e.g., senile dementia is associated with arteriosclerosis), and mental diseases have physical components (e.g., schizophrenia may be associated with genetic tendencies to the disease). There is no clear line to be drawn between mental and somatic diseases. Mental diseases are those diseases with a *predominant* mental symptomatology. Similarly, mental health identifies the psychological dimension of health without committing one to a sharp distinction, much less a separation, of mental health and physical health.

As a consequence, attempts to draw clear lines between true diseases (i.e., somatic diseases) and problems in living (i.e., mental diseases) have not been successful. In fact, such distinctions would require some sort of Cartesian dichotomy between the body as the bearer of diseases, and the mind as the perpetrator of problems in living. Not even a distinction between medical versus non-medical modes of therapy can be made along the line between somatic versus mental health. Somatic conditions can be treated as problems in living (e.g., heart disease can be viewed as a concomitant of certain life-styles of overeating, smoking, etc.) for which one is responsible and which will respond to changes in life-style, just as many mental diseases can be so viewed. *Pari passu*, both mental and physical health can be seen as elements of life-styles [26]. We are thus returned to the consideration of global meanings of the term 'mental health.' 'Mental health' remains an endorsement of one's general success in adaptation to one's environment including one's culture, while the meaning of 'success' depends upon cultural judgments about the end point of numerous physical, psychological, and social processes. As a result, 'mental health' is a very difficult term to explicate.

It is striking that most of the contributors to this volume have avoided a direct analysis of the concept of mental health. The term 'mental illness' is more often addressed here than the term 'mental health.' The reasons for that avoidance are complex and reflect in part the general difficulty with the concept of health [5, 8]. These have in part been suggested by von Wright in his discussion of medical goodness in *The Varieties of Goodness*. He claims that the primary sense of health (for the most part, von Wright is concerned with physical health) is privative in the sense of 'health' identifying the absence of pain or pain-like states, or of the frustration of the needs and wants of a 'normal life' ([24], p. 61). When we say that someone is in good health, we usually mean that he or she has nothing wrong with them ([24], p. 55). But the words 'nothing wrong' indicate that even such privative senses of health are strongly infected with value-laden judgments [8]. 'Health' in the positive sense, in the sense of 'the presence of feelings of fitness and strength and in a similar pleasant (agreeable, joyful) state' is even more value-laden ([24], p. 61). It is in this positive sense of health that 'the healthy body and mind can be said to flourish' ([24], p. 62). Again, it is health in this most ample sense which is the most difficult to analyze in any clear and straightforward fashion.

It is this and similar difficulties which led to the organization of the Symposium which culminated in this volume. The goal was to bring philosophical reflections to bear on some of the many conceptual issues raised by talk about mental health and disease. The result, the Fourth Symposium on Philosophy and Medicine, was held on May 16, 17, and 18, 1976. The scholars and scientists who participated contributed a broad spectrum of historical, conceptual, and psychiatric analyses of mental health and disease. As in past symposia, a balance has been sought among contributions from philosophers of science, analytic philosophers, and philosophers concerned with phenomenological and speculative issues. The goal has been to display the various approaches philosophy can offer. On the other hand, essays by individuals in the health care fields, as well as from historians of medicine, have been included in order to offer a view of the conceptual issues ingredient in notions of mental health as seen from the perspective of health care practitioners.

The papers of Professors Burns and Delkeskamp offer an historical portrayal of American legal tradition's reflections on mental health and illness. Chester Burns draws special attention to the conflict between the culturally accepted view of man as a rational, free, and accountable agent, and concepts of medical insanity which mitigate the sense of such rationality, freedom, and accountability. He signals, among other things, the difficulty of developing a concept of partial responsibility, and the resistance by jurists against the influence of medical concepts upon prevailing moral and legal norms. The tension between the desire to hold most persons responsible for their actions, while excusing some from responsibility under the rubric of insanity, is further explored by Corinna Delkeskamp. In particular, she indicates the strength of this dispute in the nineteenth century American legal tradition. Her article questions the extent to which being mentally healthy is the same as being a rational free agent at home in and with (and responsible for) one's thoughts and actions. The roots of our general views about mental health and disease are examined with the conclusion that these concepts have broad and deep social and legal consequences. They are not simply descriptive concepts, but are evaluative and performative as well (e.g., to say '*X* is mentally ill' is to place *X* in a special social and legal category).

The papers by Alan Donagan, Stephen Toulmin, and Daniel Creson, address our views of mental illness and the therapies it evokes. Alan Donagan questions the legitimacy of extending the concepts of disease from the body to the mind since bodily complaints are at worst sufferings, while mental diseases can alter the very character of the person afflicted. Though physical illnesses may offer an opportunity for the development of character, some have argued that a certain amount of mental illness itself contributes to one's character. In fact, Donagan argues, an adequate account of mental health and disease requires reference to such non-empirical concepts as the will. As a result, mental disease and health cannot simply be defined in terms of homeostatic function or malfunction. Instead, reference must also be made to control or loss of control over a class of actions that are usually voluntary, and/or to the possession or lack of possession of the ability to arrive at the truth about one's self and one's situation. Given this broader appreciation of mental illness, the vexations of a neurosis can be viewed as the stimulus to undertake to resolve conflicts and thus lead to a more effective self-knowledge. In short, the meaning of mental health and illness is far richer and more nuanced than that of physical health and illness.

Stephen Toulmin attempts to go behind the very debate concerning the applicability of the medical model to considerations of mental illness by examining how the 'helping professions' address the complaints of patients. He seeks to encompass both the patient-oriented and client-oriented activities of health care professionals under the more basic notion of caring for presenting complaints. Still, Toulmin wishes to preserve the distinction between those mental illnesses caused by bodily dysfunctions, and those arising more from mental life. He does this by arguing that the medical model is interpreted mechanistically, which is as inappropriate for a complete account of somatic medicine, as it is for a full account of psychotherapy. The troubles which patients tend to have are of a fabric woven out of somatic, moral, social, and legal threads of complaints. Consequently, there are reasons to prefer the language of complaint and remedy over that of disease and cure. Though problems have solutions, and diseases have cures, some complaints may simply need to be lived with and managed. They may lie beyond medical solution or cure (compare, for example, how the health professions should 'treat' a person with terminal cancer, or 'treat' someone who must adjust to a colostomy – even if there is no chance of cure, or restoration of perfect health). Moreover, mental remedies can contribute to the good of persons, even if they do not cure, for they can aid in better self-understanding or self-knowledge.

Thus, to return to an earlier point, no clear line can be drawn between the role the physician plays under the aegis of the mechanical medical model and the role he or she plays under the aegis of the mental illness model. The role of the physician is rather a broad and inclusive one, best described as that of a provider of remedies for complaints. In reply to Toulmin, Daniel Creson questions whether this amplified view of the physician's role does not border on the messianic. Further, the 'homeostatic principle' and the 'mechanistic model,' which Donagan and Toulmin disparage, remain very useful. In any event, construals of health and disease are scientific and social constructs. Creson suggests that we view Donagan and Toulmin's contributions as attempts to reshape those constructs.

Horacio Fabrega also advances an analysis of the concepts disease and health, of how they take on reality in particular cultures. He addresses disease as a symbolic category by means of a phenomenological, descriptive account of the function of medical taxonomies in aiding persons to categorize the world they experience in 'illness.' The result is a presentation of what beliefs or theories are relevant within particular cultures. In this regard, the Western de-personalized and mechanized view of the body in biomedicine is a recent and far from widespread viewpoint. In fact, individuals do not present themselves to medicine in what Fabrega terms the dualistic (i.e., separating mind and body) and associated reductionist perspectives (i.e., the somatic view of the human body devoid of its personal significance) of modern medicine. Rather, they present for medical care with personalized representations of an altered self – aches, worries, concerns, etc. This broader view appears in most cultures. Indeed, the fact that most Western concepts of disease and illness are de-personalized leads them to be socially nonsensical – they have no immediate societal significance in that they are abstracted from immediate personal concerns with illness. Concepts of psychiatric diseases have, though, maintained a close bond to appraisals of social activity. Moreover, the organ which is the focus of psychiatry's concern, the brain, underlies regulation of social activity. As a result, diagnoses of psychiatric diseases remain dependent upon the assessment of behavior. In criticism of Fabrega's account, Ruth Macklin points out that he neglects the concept of causality. Fabrega is not, so she contends, interested in the scientific accuracy of theories, but, rather, in how perspectives on health and disease function in particular societies. A greater stress on causal issues would, Macklin contends, lead to a holistic view of man as a single web of causal influences – a view which would further undermine dualistic portrayals of man and support holistic ones.

As Fabrega viewed disease as a symbolic category embedded in a particular culture, van den Berg views disease categories as bound to particular time periods. He argues that understandings of health and disease, in fact, understandings of reality, are historically bound. For example, not only was it not possible to describe neuroses as we know them (or in a fashion approximating how we know them) before George Cheyne's publication of his description of neuroses (*The English Malady*, 1733) [6], such illnesses did not exist. According to van den Berg, reality is a social and, therefore, a historical construct. Prior to the development of the requisite concepts and viewpoints, the world appeared in a fashion that did not make the experience of neuroses as we understand that term possible. It is this changing character of the experienced world that van den Berg studies under the rubric of metabletics. But, as Zaner indicates, this very metabletic view is itself historically rooted. Not only are our pictures of mental health and disease rooted in a particular historical era and its reality, but our way of understanding such history is itself historically rooted. Not only can a phenomenological account be given of mental health, but a phenomenological account of such phenomenological accounts can itself be sought. The reflections on the contributions of phenomenology to the understanding of mental health are brought to a close with Commemorative Remarks concerning the work of the late phenomenologist-psychiatrist, Erwin Straus (to which a list of his works are appended).

Speculative considerations of the nature of mental health and illness are introduced by Robert Neville, beginning with a consideration of the mind–body problem. Neville argues that a process view of mind and body, such as Whitehead's, not only resolves the legitimate points raised by Spinoza and Leibniz (i.e., for Spinoza, that a thing may be understood as having both a physical and mental nature and, for Leibniz, that a human is not a single substance but a society of monads), but also best accords with reality as we know it. By addressing these issues, Neville introduces us to the reflections of Leonard Feldstein, which are in part indebted to anticipations of process philosophy in the work of the American philosopher, Charles Sanders Peirce (1839–1914). Feldstein views man in terms of rhythmic processes for which the unconscious is the mediator and midway point between the truly bodily and the fully mental dimensions of human life. Mental health is, according to Feldstein, the integration of mental and physical activities in which the unconscious expresses itself without the contortions that pervert the rhythms of life in pathological conditions. In commenting upon Feldstein's paper, in fact in commenting generally upon such attempts in philosophy, Corinna Delkeskamp argues that the necessary conditions for the success of speculative accounts include: (1) providing a replete enough explanation in order for one to see whether ordinary descriptions of reality are benefited by a speculative portrayal of reality, and (2) showing how our ordinary understanding of reality leads us to such speculative accounts.

The papers by Irving Thalberg and Caroline Whitbeck examine the role played by ideas of freedom and compulsion in concepts of mental health and disease. Thalberg attempts an account of claims that mentally diseased persons can be described as acting against their wills. In part, he has in mind remarks such as Freud's that the role of the psychoanalyst is to give back to the neurotic's ego command over the id, the ability to act in accord with one's own intentions. In examining such issues, Thalberg provides an account of the contrast between doing what one wants and being constrained to act in a particular fashion. In particular, he gives grounds against understanding attitudinal derangements (e.g., the behavior of an addict) on the model of coercion (i.e., being compelled by a desire rooted in an addiction is in important respects unlike being compelled by a mugger to part with one's money so that the usual hierarchical and non-hierarchical analyses of coercion fail to account for attitudinal disturbances). Caroline Whitbeck argues that while alcoholism and drug addiction do not answer to Thalberg's description of motivational disturbances, neuroses do. When impulses are experienced as ego-alien, they may be properly described as coercive (i.e., as something other than we bringing force to bear upon us). But in the case of psychoses, the disruption of mental life may be so severe as to preclude talk about anyone acting, and, therefore, of anyone being coerced. Further, insofar as self-deception erodes mental health, one will need hierarchical analyses of self-deception (i.e., analyses which distinguish between (i) the level at which one wants to satisfy an inclination, and (ii) the level at which one does not wish to admit having the desire to satisfy the inclination in question).

The issue of freedom appears in quite a different fashion in the essay by Thomas Szasz on the concept of mental illness. Szasz argues that 'mental illness' is a metaphor, that it indicates a social role, rather than a pathological condition. Moreover, 'mental illness' is a prescriptive, not a descriptive, term, so Szasz argues, as is not the case with true illnesses – somatic illnesses. To call someone mentally ill is to place him in the social category, mental patient, not to explain a pathological condition. Also, the social and legal distinction between the mentally ill and mentally healthy requires a quasi-causal distinction between states caused by pathological processes, and those initiated by free will. This distinction, so Szasz argues, conflates giving an explanation by accounting for a causal chain with giving an explanation by supplying a hidden motive. Psychiatrists (at best) do the second, so Szasz argues, while holding that they are scientists engaged in a process tantamount to doing the first. Psychiatrists give justifications for social labeling under the guise of giving causal explanations. In commentary, Baruch Brody argues that Szasz, at most, establishes that current concepts of mental illness fail, not that all such concepts fail in principle. Diagnoses of mental illness are, so Brody contends, at least in part based on publicly observable behavior, and are thus open to confirmation or disconfirmation in a manner not essentially different from many somatic diseases. In addition, many identifications of deviations from biological norms presuppose value-judgments. The clear distinction which Szasz wishes to draw between descriptive and prescriptive concepts, between somatic and mental concepts of illness, according to Brody, fails.

Finally, a series of general critiques of the papers presented in this volume is offered through the Round Table Discussion. There, Fabrega addresses van den Berg's concept of the time-bound nature of concepts of disease, as well as Szasz's assertion of the non-objective status of diagnoses of mental disease. Picking up the latter theme, James Knight indicates how Freud developed his notions of mental health and illness by assimilating the concepts of somatic health and disease, as well as the concept of happiness, producing a mixed model of mental health and illness. Much of the dispute concerning concepts of mental health and illness turns, so Knight contends, on decisions about whether to seek a unitary concept of health, or to assign the treatment of somatic disease to a physician and problems of living to a non-medical caregiver. Multi-variable views of health and disease will favor a unitary, more holistic, view of human pathologies, though the issues of free will versus determinism and of political misuse of medical concepts will remain.

In contrast, Karen Lebacqz and Bernard Towers raise issues in a more personal fashion. Lebacqz gives a case history of a delusional person and asks how one should judge behavior to be normal or abnormal, and how best to seek 'help' for such persons (and how to decide what 'help' is appropriate). Bernard Towers suggests that the issues at stake in understanding mental health are best addressed on the basis of the work one does on one's analyst's couch. Moreover, one would understand the essays in this volume better if one had some knowledge of the personal viewpoints and orientations of the authors. The discussion is then closed by Professors Thomas Szasz and Edmund Erde. Erde indicates the complexity of the semantics of health which perform evaluative, prescriptive, and descriptive functions. He suggests that mental health and illness themselves may turn on basic paradigms about the nature of reality, though mental health may have as a necessary feature a certain level of flexibility and ability to recognize the universal character of reasons. Mental illness becomes the inability to gear into a common world, the paradigm of everyday life.

We are brought, then, to the point at which this introduction began – concepts of mental health and illness turn on basic judgments concerning reality and adequate participation or recognition of reality. One can see, then, both the interest of philosophers for issues in mental health, and the contributions that philosophy can make. Talk about mental health presupposes judgments concerning the appropriateness of distinctions between mind and body, and between caused behavior and freely chosen actions. Discussions of mental health raise epistemological issues concerning whether the nature of reality is discovered or in some sense fashioned as an intersubjective construct. The papers in this volume, as a result, range from analyses of ethical issues to considerations of problems in the theory of action, in the philosophy of mind, and in epistemology.

I wish to express the gratitude of the editors to the Hogg Foundation for Mental Health for its support of the Symposium, 'Mental Health: Philosophical Perspectives,' at which the original versions of these papers were presented, and to the Institute for the Medical Humanities of the University of Texas Medical Branch which hosted this Symposium and aided in its organization. Special thanks are due, as well, to Ernest S. Barratt, William B. Bean, Marjorie Huffman, Judge Jerome Jones, Jodie Leecraft, William C. Levin, Irwin C. Lieb, John P. Vanderpool, and Robert B. White, who in various and very important ways contributed to the success of the Symposium, and, thus, indirectly to this volume. Finally, I wish to express my thanks to Edmund Erde, John Moskop, Ruth Walker Moskop, Jane Backlund, and Stephen Wear, who worked with care and diligence in the preparation of the manuscript of this volume.

Galveston
December 21, 1976

H. TRISTRAM ENGELHARDT, JR.

BIBLIOGRAPHY

1. Arieti, Silvano: 1955, *Interpretation of Schizophrenia*, Robert Bruner, New York.
2. Bleuler, Eugen: 1950, *Dementia Praecox or The Group of Schizophrenias*, International Universities Press, New York.
3. Bleuler, Eugen: 1936, *Textbook of Psychiatry*, The Macmillan Company, New York.
4. Breur, Josef and Freud, Sigmund: 1955, 'Studies on Hysteria', in *The Complete Psychological Works of Sigmund Freud*, vol. II, Hogarth Press, London, p. 305.
5. Burns, Chester R.: 1975, 'Diseases Versus Healths: Some Legacies in the Philosophies of Modern Medical Science', in H. Tristram Engelhardt, Jr. and Stuart F. Spicker (eds.), *Evaluation and Explanation in the Biomedical Sciences*, D. Reidel Publishing Company, Dordrecht and Boston.
6. Cheyne, George: 1733, *The English Malady: or, A Treatise of Nervous Difeafes of All Kinds*, G. Strahan, London.
7. Constitution of the World Health Organization (preamble): 1958, *The First Ten Years of the World Health Organization*, World Health Organization, Geneva, p. 459.
8. Engelhardt, H. Tristram, Jr.: 1976, 'Human Well-being and Medicine: Some Basic Value-Judgments in the Biomedical Sciences', in H. Tristram Engelhardt, Jr. and Daniel Callahan (eds.), *Science Ethics and Medicine*, The Hastings Center, Hastings-on-Hudson, New York.
9. Engelhardt, H. Tristram, Jr.: 1976, 'Ideology and Etiology', *The Journal of Medicine and Philosophy* **1**, 256–268.
Chicago.
10. Engelhardt, H. Tristram, Jr.: 1973, *Mind-Body: A Categorial Relation*, Martinus Nijhoff, The Hague.
11. Engelhardt, H. Tristram, Jr.: 1975, 'The Concepts of Health and Disease', in H. Tristram Engelhardt, Jr. and Stuart F. Spicker (eds.), *Evaluation and Explanation in the Biomedical Sciences*, D. Reidel Publishing Company, Dordrecht and Boston.
12. Freud, Sigmund: 1961, 'The Ego and the Id', in *The Complete Psychological Works of Sigmund Freud*, vol. XIX, Hogarth Press, London, p. 50n.
13. Freud, Sigmund: 1964, *The Standard Edition of the Complete Psychological Works of Sigmund Freud*, vol. XXIII, Hogarth Press, London, pp. 144f.
14. Friedman, Meyer and Rosenman, Ray H.: 1959, 'Association of Specific Overt Behavior Pattern with Blood and Cardiovascular Findings', *Journal of the American Medical Association* **169**, 1286–1296.
15. Heath, Robert G.: 1960, 'A Biochemical Hypothesis on the Etiology of Schizophrenia', in D. D. Jackson (ed.), *The Etiology of Schizophrenia*, Basic Books, New York, p. 146.
16. Holmes, Thomas H. and Rahe, Richard H.: 1967, 'The Social Readjustment Rating Scale', *Journal of Psychosomatic Research* **11**, 213–218.
17. Lazare, Aaron: 1973, 'Hidden Conceptual Models in Clinical Psychiatry', *The New England Journal of Medicine* **288**, 345–351.
18. Margolis, Joseph: 1976, 'The Concept of Disease', *The Journal of Medicine and Philosophy* **1**, 238–255.
19. Sullivan, Harry Stack: 1953, *The Interpersonal Theory of Psychiatry*, W. W. Norton & Co., New York, p. 18.
20. Szasz, Thomas: 1974, *Myth of Mental Illness: Foundations of a Theory of Personal Conduct*, Harper & Row, New York.

21. Szasz, Thomas: 1965, *The Ethics of Psychoanalysis*, Basic Books, New York, p. 18.
22. Tuke, D. Hack: 1894, 'Imperative Ideas', *Brain* **17**.
23. von Domarus, Eilhard: 1925, 'Über die Beziehung des normalen zum schizophrenen Denken', *Archiv für Psychiatrie und Nervenkrankheiten* **74**, 641–646.
24. von Wright, Georg Henrik: 1963, *The Varieties of Goodness*, The Humanities Press, New York.
25. Wittgenstein, Ludwig: 1963, *Philosophical Investigations*, trans. G. E. M. Anscombe, Basil Blackwell, Oxford, No. 255.
26. Wolf, Stewart: 1961, 'Disease as a Way of Life: Neural Integration in Systemic Pathology', *Perspectives in Biology and Medicine* **4**, 288–305.

SECTION I

AMERICAN LEGAL PERSPECTIVES ON INSANITY: SOME ROOTS IN THE NINETEENTH CENTURY

CHESTER R. BURNS

AMERICAN MEDICO-LEGAL TRADITIONS AND CONCEPTS OF MENTAL HEALTH: THE NINETEENTH CENTURY

For the study of historical relationships between law and psychiatry, many approaches are possible. A few scholars have focused on particular relationships [22, 19, 23], while most have incorporated legal aspects into more general reviews of the history of psychiatry in the United States[1] [11, 13, 8]. Yet to be explored in any depth are the legal implications of moral therapy, legal aspects of the mental hygiene movement as it emerged during the latter part of the 19th century, and legal terms used as metaphors by physicians characterizing concepts of health before 1900.[2] Another approach would be to examine the thoughts of American physicians who wanted their 19th century colleagues to accept the importance of utilizing their knowledge of insanity in helping courts of law perform their societal tasks. Although more narrow in scope and seemingly more remote from a study of mental health concepts, I believe that this focus of historical inquiry can be illuminating.

American physicians inherited centuries of concern about the legal aspects of medicine. These traditions extended as far back as the Roman Empire and certainly included activities at the civil and ecclesiastical courts evolving during the Middle Ages. In France, especially after 1200; in the Italian city states, especially after 1400; and in the Germanic states, especially after 1500, physicians and surgeons were summoned to give testimony at courts about cases of abortion, murder, poisoning, suicide, and bodily injury. During the 16th and 17th centuries, Continental physicians such as Paré, Weyer, Augenius, Cardanus, Fidelis, and Zacchias prepared treatises about forensic problems associated with medical practice, including those pertaining to the jurisprudence of insanity. Between 1787 and 1821 a few British physicians – Samuel Farr, Thomas Percival, George Edward Male, and John Gordon Smith – published important analyses of medical jurisprudence. During the 19th century, American physicians self-consciously transformed these Continental and British heritages into an American tradition. The jurisprudence of insanity was a central feature of this American legacy.[3]

Contrary to some opinions of today, outstanding American physicians acknowledged the importance of civic obligations long before Medicare and Medicaid were adopted. This acknowledgment of professional responsibilities to a community was even included in a separate chapter of the code of ethics

H. T. Engelhardt, Jr. and S. F. Spicker (eds.), Mental Health: Philosophical Perspectives, 3–14.

adopted by the American Medical Association in 1847. In addition to other duties, physicians were expected to utilize their medical knowledge as needed by coroners, lawyers, and judges attempting to handle legal problems of the American society.

Almost four decades earlier, Benjamin Rush had discussed the jurisprudence of insanity in what was to become the first printed essay on medical jurisprudence written by an American clinician ([20], pp. 363–395). Theodric Romeyn Beck had included several chapters on insanity in the first American textbook on medical jurisprudence published in 1823 [2]; and, in 1838, Isaac Ray had published the first monograph on the jurisprudence of insanity written in the English language [18]. Heralded by Ray's efforts, 19th century American physicians became recognized internationally for their attention to the legal aspects of insanity.

From the time of Benjamin Rush in the early part of the century to that of Shobal Clevenger at the end of the century, the primary obligation of a physician to the American courts was that of testifying as an expert witness.[4] A witness was limited to the facts of a case. The factual problem, in most situations, was to determine whether or not insanity existed. If it did, then the court and jury had the problem of determining the extent to which this insanity excused one from social obligations or legal punishment.

Physicians, strictly speaking, had only one task in these courtroom situations, that of diagnosis. To use some of Rush's categories, did the defendant have mania, or amenomania, or manalgia, or a derangement of the will, the memory, the passions, or the moral faculties? Give us a diagnosis, Dr. Rush, and let the lawyers and laymen do the rest; do not inject your moral or legal opinions.

But it was not that simple. Experts could also give opinions and the dividing lines between the facts of insanity and the legal consequences of those states were not very sharp. Try as hard as they might, many physicians could not separate their diagnostic labels from the legal and social consequences of those labels.

Some forms of irrational behavior did not exempt a person from legal liability, claimed Rush. These included those suffering from partial madness, in which the mind was deranged on one subject only; or those with unusual mental excitement, but not mania; or those who were demented or dumb as long as they understood the purpose of money; or those experiencing a partial loss of memory. Some forms of mental illness – general mania, or delirium, or 'where persons depart in their feelings, conversation, and conduct, in a great degree from their former habits' ([20], p. 368) – incapacitated persons so they could not make valid wills nor dispose of property nor serve as witnesses in courts. Finally, Rush claimed that certain criminal acts were done by those who were experiencing an insanity of the will or a derangement of a moral faculty. He believed that acts of murder or theft performed by these persons were the effects of passions involuntarily directing actions without a derangement of the person's intellect. These mentally ill persons should be exempted from any punishment ordinarily rendered to such criminals. Rush even hoped that the new medical insights about the human mind would lead to the abandonment of capital punishment altogether.

Thus, at the very outset of American interest in the legal aspects of insanity, physicians saddled themselves not only with the problem of determining the existence of insanity, but the added problem of associating these diagnostic efforts with the legal and social consequences of their professional decisions. The burdens became rather awesome.

Conscientious practitioners had to distinguish insanity from other forms of illness: febrile delirium, hypochondriasis, simple hallucination, epilepsy, delirium tremens, and senile dementia. Practitioners then had to differentiate different forms of insanity, deciding whether dementia or melancholia existed; and whether or not it existed in a total or partial state. Added to these diagnostic difficulties was the realization that some diagnoses could have extraordinary legal and social consequences.

Partial insanity might exempt in civil cases, but not in criminal ones. Persons who committed criminal acts during a 'lucid interval' could be punished, but those committing criminal acts during the smogs and fogs of irrationality would not be punished. Furthermore, physicians had to distinguish actual insanity from pretended insanity, as successful pretense would allow a criminal to avoid punishment. And, if it were not already complicated enough, some physicians and lawyers believed that unimpaired moral judgment (or the capacity to distinguish right from wrong) could coexist with the intellectual derangement usually defined as insanity.

Maybe I have been too generous in claiming that 19th century American physicians were interested in diagnosing insanity, but I think not. Surely they wanted labels for diseases, descriptions of abnormal behavioral states, definitions of irrationality. I believe most would have denied that they were attempting to characterize something other than disease. What, then, does all of this have to do with concepts of mental health?

As I have argued previously, the creation of disease labels or nosologizing was one of the most important professional pastimes of 19th century physicians [6]. Lurking behind these nosological quests was a notion of health as the logical opposite of disease. We might assume that attempts to characterize insanity would be based on some basic assumptions about sanity. A concept of sanity or mental health would provide some criteria by which to characterize deviation or insanity and some goal by which to evaluate or justify therapeutic and hygienic efforts. In other words, if you do not know what sanity is, how can you define insanity; and if you do not know what sanity is, how can you know what to do to sustain or restore it?

Can we find any clues to concepts of sanity or mental health in the 19th century attempts to develop the jurisprudence of insanity in the United States? Recall that Rush was willing to exempt from civil obligations those persons who 'depart in their feelings, conversation, and conduct, in a great degree from their former habits,' suggesting that a person who had his thoughts, feelings, and actions well ordered in habitual routines, was enjoying a healthy mind, was sane. We breathe expectantly as we encounter Beck's thoughts about a 'state of mind necessary to constitute a valid will.'[5] Yet, with the exception of some comments about age and sex, Beck does not characterize the nature of a mind adequate for preparing a valid will. Rather, he examines numerous conditions and diseases that incapacitate persons from making such wills.

Quoting from a work by Andrew Combe, Isaac Ray defines insanity as 'the prolonged departure, without an adequate external cause, from the state of feeling and modes of thinking usual to the individual when in health . . .' ([18], p. 111). Ray, unlike Rush and Beck, uses the word 'health' and we expect Ray to tell us something about his idea of mental health. We marvel when he claims that concepts of mental health are related to metaphysics and the philosophy of mind and concludes that one cannot explain health or disease on the basis of metaphysical speculation. He also berates metaphysicians who ignore problems of disease ([18], pp. 55–57).

Here I am, then, in a symposium on philosophy and medicine telling the audience that the outstanding physicians fashioning the jurisprudence of insanity in 19th century America were apparently uninterested in concepts of mental health and definitely uninterested in the intellectual strivings of moral and mental philosophers. Was it really true, and if so, what do we make of it?

Rush recommended the study of Christian ideals and of metaphysics, but not moral philosophy [7]. He believed that moral philosophy had no place in education, being pagan and therefore anti-Christian. Metaphysics was neither classical nor modern philosophy. Rush defined metaphysics as 'a simple history of the faculties and operations of the mind, unconnected with the ancient nomenclature of words and phrases, which once constituted the science of metaphysics.' Since the moral faculties were important components of the mind, Rush encouraged their investigation, but not at the expense of Christian principles. This value judgment did not disappear as the century evolved.

Others, like Isaac Ray, believed that physicians studied metaphysics 'too much instead of too little.' Added Ray, 'Metaphysics in its present condition is utterly incompetent to furnish a satisfactory explanation of the phenomena of insanity, and a more deplorable waste of ingenuity can hardly be imagined than is witnessed in the modern attempts to reconcile the facts of one with the speculations of the other.' When John Ordronaux summarized the legal precedents constituting the jurisprudence of insanity by 1878, his first rule was that the constitution of the mind was unknowable. 'The law can take no cognizance of the human mind dissevered from a living body, nor of its operations in a state different from that belonging to such an existence. It deals alone with the finite. It can only appeal to universal consciousness and rest there. Consequently, the doctrines of metaphysics cannot be accepted by courts as guides for the elucidation of states of mental disorder. From Aristotle to Sir Wm. Hamilton, mankind however reasoning high

> Of Providence, fore-knowledge, will and fate,
> Fixed fate, free-will, fore-knowledge absolute;

are yet no more advanced today than were Milton's fallen angels, who, after discussing these problems in solemn conclave, retired discomfited,

> And found no end, in wandering mazes lost ([17], p. xxv).

So Rush, Ray, and Ordronaux had little use for the professional philosophizing of their day and there were few, if any, physicians or alienists attempting to characterize a concept of mental health as a prerequisite for defining mental illness, either for purposes of diagnosis and treatment or for purposes of providing expert testimony in judicial proceedings ([11], pp. 63, 185–186).

For some time now, I have believed that these doctors were, for the most part, consciously rejecting the speculative traditions of their Western culture; that the empiricism and positivism of 19th century science and the pragmatism and utilitarianism of bedside practice had caused American practitioners to eschew intentionally the paradises of Western philosophers. My foray into the jurisprudence of insanity has led me to a reassessment of this claim and I offer another interpretation, which may seem paradoxical.

Although conciously opposing philosophical speculation about the nature of the human mind, the majority of American physicians caring for the mentally ill had very definite concepts of mental health, concepts that were rooted deeply in Puritan religious traditions and concepts that found considerable expression in the patterns of moral philosophy taught and studied in American colleges and universities throughout the 19th century. The sane person, the mentally healthy person, was rational, free, responsible. He was rational and reasonable; he had the capacity to choose freely his actions and responses; he was accountable and responsible to the laws of man and God and he could be punished for breaking those laws. It was not necessary for physicians to write books and essays about the fundamental assumptions of their culture. They were basic premises which had been learned in homes and churches and schools, especially during the last years of their collegiate education. Remember that most of the men included in this study of the medical jurisprudence of insanity were among the minority of well-educated 19th century American physicians.

As Wilson Smith has demonstrated, the moral philosophy of that century was a curious blend of law, religion, philosophy, history, government economics – a melding of what we would call today the humanities, behavioral sciences, and social sciences. Most, if not all, of these areas were included in particular moral philosophy texts usually studied during the senior year of a college education. Although there was variation in the ethical pronouncements of these books, common assumptions were characteristic. These texts offered 'a universe of moral laws in which the uncertainties of life could be met with theologically satisfying preconceptions and fixed plans of action' ([21], p. 28). In spite of variant shades of religious orthodoxy, this moral philosophy championed the rationality of man, his capacity for freely controlling his actions, and his omnipresent duties and obligations to himself, his fellow men, and his God. Even though the 'theoretical' ethics of this moral philosophy were rooted in theology, the 'practical' ethics typically considered various aspects of jurisprudence such as concepts of justice and the obligations of judges and legislators. This moral philosphy, during the 19th century, was a basic part of education for the legal profession, and there was remarkable congruence between this moral philosophy and the philosophical bases of American law.

Understanding this congruence enables us to gain a better appreciation of 19th century relationships between law and psychiatry, and specifically the legal ramifications of a diagnostic label: moral insanity. As long as physicians defined insanity as the loss of reason, then the accompanying loss of free will and responsibility to the laws of man and God was a logical consequence within the framework of the fundamental assumptions of the American culture. But the claim by some physicians that persons could become insane and still have their reasoning powers intact was a threat not only to the set of moral values predominating, but to the very basis of law itself.

Rush, Ray, and others did not confine their professional tasks to that of diagnosing a form of emotional or affective or moral insanity. They went much further and claimed that even though a person understood the moral consequences of his criminal act, he was still sick and should be exempted from legal punishment. Not only did Rush and Ray involve themselves in the politics of conceptual reform in law and moral philosophy; they also attempted to modify the commonly accepted model of human nature with a concept of mental health that allowed sanity to coexist with uncontrollable passion. Using the language of disease and insanity, they intruded into the generally accepted domain of legal and moral prerogatives. It is not surprising that their colleagues – medical and legal – could not tolerate this intrusion.

There was a concept of mental health tenaciously embedded in the legal and moral assumptions of the day and it is no wonder that physicians had difficulty separating their diagnostic judgments from their moral ones. It is also no wonder that most lawyers and judges of the day would not allow physicians to use the language of disease as a method of undermining the concept of mental health already sanctioned by legal, religious, and educational institutions. As one judge said '. . . that cannot be health in law which is disease in fact . . .' ([9], p. 22). If there were mental diseases in fact, and most lawyers and physicians thought there were, then the law should not champion an ideal of mental health that was counter to these disease entities. In short, true mental disease would exempt from legal responsibility. But if there was no disease in fact, then the concept of mental health incorporated in the law was that same set of values which held all men responsible to the laws of man and God. No one state of disease like moral insanity – real or imagined – could remove this responsibility.

Few were the 19th century physicians, lawyers, and judges who acknowledged a concept of moral insanity which exempted from legal punishment.[6] Charles Coppens, a Jesuit priest who taught medical ethics at Creighton Medical College in Omaha around the turn of the 20th century, captured the spirit of the day. Coppens objected to the notion in insanity defined as a perversion of the will which allowed a person to know the difference between right and wrong and yet be unable to do what was right. 'It is against those clear principles of psychology and ethics which are not only speculatively evident, but practically necessary to maintain the fabric of human society' ([10], p. 186). Coppens believed that moral depravity was often exhibited before the courts as so-called moral insanity. He claimed that a sound whipping would often stop the nuisances of the morally depraved. 'The rod for the juvenile offender, and the whipping post for adults, would cure many a moral leper and be a strong protection for society at large, especially if applied before bad habits freely indulged have demoralized the person beyond the usual limits' ([10], p. 195). (Remember that these were exhortations to medical students included in a lecture on professional ethics.)

American physicians creating the jurisprudence of insanity had other concerns. In addition to problems of diagnosis and problems of confounding diagnostic judgment with moral opinion, physicians were disturbed by other facets of the legal system. Foremost was the conceptual escape hatch which lawyers and judges had fashioned with the term 'unsoundness of mind.' Persons could be medically sane, but legally exempt; this might include idiots, senile persons, and habitual drunkards. Legally, in fact, a person could be neither insane nor mentally deficient and still be free of legal responsibilities. Secondly, the law would not measure degrees of weakness in one who was mentally healthy, but the law would admit different kinds of insanity and, accordingly, varying legal consequences. In the third place, the law would not allow someone to be totally mad. Reason and unreason were constantly juxtaposed in the mentally ill. 'No lunatick was wholly without reason' ([17], p. xxxi). Whether in justification of the therapeutic goals of medicine or in justification of the ideal of a morally free agent, the law recognized both partial insanity and lucid intervals. 'And as nature always works toward restoration, so gleams of reason are ever bursting through the clouds of mental darkness' ([17], p. xxxii). Yet these gleams of reason were not self-complete, but only limited degrees of brightness which enabled 'the party to judge soundly of the act' ([17], p. xxxvi). Such splitting of fine legal hairs only added more confusion to an already impossible situation.

What had started in the beginning of the 19th century as an effort on the part of American physicians to fulfill new civic obligations, had turned into a conceptual nightmare for physicians, lawyers, and judges. The problem of characterizing a host of mental diseases was challenge enough. 'There indeed seems to be a name for every conceivable kind of mania, except that of maniacal classification, or insane nomenclature,' said one mid-century physician-lawyer ([12], p. 355). But the problem for physicians as expert witnesses was not simply that of offering a label. As experts, they could give opinions as well as describe facts. It was extremely difficult for the physicians of the day to separate their diagnostic claims from their moral and legal opinions; to arrive at an objective characterization of a criminal as insane, free of any legal or moral sentiments. Compounding the difficulties of the situation were the tortuosities of legal terms and procedures and the tendencies of some lawyers and judges to expose the contradictions and inconsistencies of medical testimony, thereby discrediting physicians and the idea of an expert witness. It was, after all, a rather easy task to accomplish.

The temper of the latter part of the 19th century was captured in a comment made by a lawyer-physician, John J. Elwell. 'Neither court or counsel can feel the want of settled principles more than the medical witness. Complexity, contradiction, difficulty, doubt and obscurity are the rules, if many be so called, that guide him! These alone are certain, and present in every case' ([12], p. 378). What can we conclude from this excursion into 19th century relationships between law and psychiatry?

First, physicians were interested primarily in disease. For the most part, they believed that insanity was a disease of the body and that true insanity would lead to a loss of the fundamental characteristic of human nature: reason. If true insanity existed, then this loss of reason resulted in a loss of free will and a loss of accountability to God and man. This notion of mental disease as a physical disorder leading to a derangement of the intellect was more or less accepted by the courts. 'Insanity at law means a permanently disordered state of mind beyond the control of the individual, and produced by disease' ([17], p. xxvii). The problem for physicians was to characterize the nature of mental disease. This objective was most acceptable because the importance of clinical diagnosis was being championed throughout the profession of medicine.

In the second place, American physicians of the 19th century were confronted with new social obligations, a need to utilize medical knowledge in responding to the legal problem of their communities and not exclusively for the purpose of caring for individual patients. Those caring for the so-called mentally ill were expected to contribute their medical knowledge as expert witnesses when courts were called upon to adjudicate criminal or civil problems involving those alleged to be mentally ill. But because of their diagnostic difficulties, expertness was not the order of the day. Even though the nosologies of mental illness were quite unsatisfactory, courts began to accept notions of insanity as justification for exempting persons from social liability and legal punishment. The boundary between diagnostic labeling and moral evaluation was blurred indeed. When physicians like Rush and Ray claimed that some criminals were insane even though their rational faculties were intact, the moral and legal order of the American universe was threatened with destruction. Consequently, credibility of physicians as expert witnesses about problems of the jurisprudence of insanity was questioned and an increasingly conservative spirit prevailed during the latter half of the 19th century. The concept of mental health contained in the moral and legal philosophy of the period was supreme and was not to be changed by the ideals of a few reformist physicians.

In the third place, physicians caring for the mentally sick seemed to be uninterested in developing a basic concept of mental health. They objected to the speculations of mental and moral philosophers and during the latter decades of the 19th century they carried on their policies of custodial care without much interest in the budding social and behavioral sciences [15]. Until recently, I thought this was a manifestation of their lack of education and of their essentially anti-philosophical posture. But I now suggest that 19th century American physicians concerned with the care of the mentally ill and expected to give testimony as expert witnesses were actually bound rather adhesively to the cultural order to their century. Their philosophy of man as a rational, free, and accountable agent was so pervasive that they would not allow some of their own professional colleagues to undermine this order even with cherished labels of new diseases. If this be true, then it is not surprising that medical practitioners would not diagnose insanity in a host of judicial proceedings ([11], p. 115). And it is not surprising that juries would frequently ignore the testimony of psychiatrists as they arrived at their legal decisions.[7]

University of Texas Medical Branch,
Galveston, Texas.

NOTES

[1] For an analysis of historiography in psychiatry, see Mora [16].

[2] One aspect of the quest for 'laws' of health will be analyzed in my article, 'The Non-Naturals: A Paradox in the Western Concept of Health' [4].

[3] For background information, see the articles by Ackerknecht [1] and my introduction to the reprint of Percival's *Medical Ethics* [5].

[4] The need to become expert at witnessing motivated the delivery of numerous lectures in medical schools and the writing of many essays and books on medical jurisprudence, especially the jurisprudence of insanity. Important American authors other than those mentioned in the text were Thomas Simons, Charles Johnson, C. B. Coventry, Charles Lee, Thomas Buckham, William Hammond, John Gray, Theodore Fisher, Clark Bell, George Beard, H. E. Allison and Edward Mann. For their writings, see Brittain [3].

[5] I looked at several editions of Beck [2]; see, for example, the sixth edition, 1838, vol. 1, p. 642.

[6] For additional information, see Halleck [14].

[7] Probably the most outstanding example of the century is that of the trial of Garfield's assassin, Charles J. Guiteau [19].

BIBLIOGRAPHY

1. Ackerknecht, Erwin H.: 1950–51, Various Articles, *Ciba Symposia* **11**, 1286–1304.
2. Beck, Theodric Romeyn: 1823, *Elements of Medical Jurisprudence*, 2 vols., Albany.
3. Brittain, Robert P.: 1962, *Bibliography of Medico-Legal Works in English*, Fred B. Rothman, South Hackensack, New Jersey.
4. Burns, Chester R.: 1976, 'The Non-Naturals: A Paradox in the Western Concept of Health', *Journal of Medicine and Philosophy* **1**, 202–211.
5. Burns, Chester R.: 1975, Introduction to reprint of Thomas Percival's *Medical Ethics*, Robert E. Krieger Pub. Co., Huntingdon, New York, pp. xiii–xxviii.
6. Burns, Chester R.: 1975, 'Diseases Versus Healths: Some Legacies in the Philosophies of Modern Medical Science', in H. Tristram Engelhardt, Jr. and Stuart F. Spicker (eds.), *Evaluation and Explanation in the Biomedical Sciences*, D. Reidel Publishing Company, Dordrecht and Boston, pp. 29–47.
7. Burns, Chester R.: 1976, 'Opening Remarks', in H. Tristram Engelhardt, Jr. and Stuart F. Spicker (eds.), *Philosophical Medical Ethics: Its Nature and Significance*, D. Reidel Publishing Company, Dordrecht and Boston, pp. 21–26.
8. Caplan, Ruth B.: 1969, *Psychiatry and the Community in Nineteenth Century America*, Basic Books, New York.
9. Chaille, Stanford E.: 1876, *Origin and Progress of Medical Jurisprudence 1776–1876*, Philadelphia.
10. Coppens, Charles: 1897, *Moral Principles and Medical Practice: The Basis of Medical Jurisprudence*, 1st ed., Benziger, New York.
11. Dain, Norman: 1964, *Concepts of Insanity in the United States, 1789–1865*, Rutgers University Press, New Brunswick, New Jersey.
12. Elwell, John J.: 1860, *A Medico-Legal Treatise on Malpractice and Medical Evidence, Comprising the Elements of Medical Jurisprudence*, New York.
13. Grob, Gerald: 1973, *Mental Institutions in America: Social Policy to 1875*, Free Press, New York.
14. Halleck, Seymour: 1968, 'American Psychiatry and the Criminal: A Historical Review', *International Journal of Psychiatry* **6**, 185–208.
15. Marx, Otto: 1968, 'American Psychiatry without William James', *Bulletin of the History of Medicine* **42**, 52–61.
16. Mora, George: 1965, 'The History of Psychiatry: A Cultural and Bibliographical Survey', *Psychoanalytic Review* **52**, 154–184.
17. Ordronaux, John: 1878, *Commentaries on the Lunacy Laws of New York, and on the Judicial Aspects of Insanity at Common Law and Inequity, Including Procedure, as Expounded in England and the United States*, John D. Parsons, Jr., Albany.
18. Ray, Isaac: 1838, *A Treatise on the Medical Jurisprudence of Insanity*, Boston. New edition, 1962, Winfred Overholser (ed.), Belknap Press of Harvard University Press, Cambridge.
19. Rosenberg, Charles E.: 1968, *The Trial of the Assassin Guiteau: Psychiatry and Law in the Gilded Age*, University of Chicago Press, Chicago.
20. Rush, Benjamin: 1811, *Sixteen Introductory Lectures Upon the Institutes and Practice of Medicine, With a Syllabus of the Latter*, Bradford and Innskeep, Philadelphia.
21. Smith, Wilson: 1956, *Professors and Public Ethics*, Cornell University Press, Ithaca, New York.

22. Szasz, Thomas: 1963, *Law, Liberty and Psychiatry,* Macmillan, New York.
23. Zilboorg, Gregory: 1944, 'Legal Aspects of Psychiatry', in J. K. Hall (ed.), *One Hundred Years of American Psychiatry,* Columbia University Press, New York, pp. 507–584.

CORINNA DELKESKAMP

PHILOSOPHICAL REFLECTIONS IN THE NINETEENTH CENTURY MEDICOLEGAL DISCUSSION

A Symposium on Medicine and Philosophy is *a priori* a suspicious enterprise. Even though certain problems in the medical field have been diagnosed as beyond the narrow limits of the discipline (strictly understood), philosophers ought to weigh prudently those cuckoo's eggs before adopting them as their own. Interest in medicine is no excuse for philosophizing about it; nor should the recent graciousness of our medical friends in asking for our opinions induce us to imagine that it is our business. Indeed, the 19th century medicolegal discussion of insanity provides ample illustration of a healthy hostility – wholeheartedly agreed upon by both the medical and legal factions – against metaphysical quibblers meddling with an affair which they rightly felt they could settle for themselves.[1] This resentment was matched by a remarkable lack of interest on the part of academic philosophers.

Looking back at that issue today, we can conceive its philosophical relevance in two ways. On the one hand, we might argue that then, just as now, the medical (or legal) issue of mental health related to problems such as the unity of mind, moral responsibility, and rational choice which properly belong to the domain of philosophy. Allowing for a sufficiently tolerant view of what constitutes philosophical awareness, one might be willing to discover 'some philosophizing or other' in all realms of human existence and, thus, with a broader definition of 'philosophical problem,' find even in medicine (or in the law) a reason for philosophical concern. The need in those fields for clarification of implicit assumptions and for criteria for deciding value conflicts is then to be taken as a call for the philosopher's help. He, as the expert, may deliver solutions.

On the other hand, we might consider such an assumption of universal competence on the part of philosophers unwarranted. It might be held that philosophy does not find its problems by looking abroad but rather invents them. After all, the philosopher, thus understood, deals not in facts but with concepts. While the latter are comprehended reflexively and in transcendental aloofness, the former, unless they yield to conceptual penetration, may be abandoned to indifference. The medicolegal discussion then appears as merely a mechanical endeavor to adjust vocabularies which had – in each of the participating disciplines – developed in opposite directions, and subsequently

H. T. Engelhardt, Jr. and S. F. Spicker (eds.), Mental Health: Philosophical Perspectives, 15–38.

to define new rules for practical decisions on confinement or execution, acquittal or correction. We might argue that this realignment of medical and legal terminologies, far from eliciting philosophically interesting puzzles, was effected even without rational communication or mutual understanding ever taking place. Nobody seemed to engage his opponent's arguments or to become aware of his own conceptual limitations when trying to refute an opponent. A gigantic battle in books and courts, by shuffling one language game through another, must have mysteriously fermented these games into useful maxims while secret sympathies among the humors of semantic particles coalesced them into sensible definitions. Thus only a complete withdrawal into its own abstracted empyrean could save philosophy from the humiliating conclusion that the absence of reason and symposia, in this case, did little to hinder the progress of civilized humanity.

Neither of these accounts of the role of philosophy vis-à-vis the medico-legal discussion is desirable. Is there, then, a third interpretation which would acknowledge that issues fall under the rubric of other disciplines without the 'primary discipline' relinquishing all concern? Can philosophy relate – say – to medicine and yet resist the temptation to solve the medical problems as if they were its own, thereby assuming a responsibility it is not fit to bear? Can medicine afford the patience to endure a dialogue where no useful answers are to be expected? Let us suppose that interdisciplinary discussion makes sense at least when it affects the manner in which the participating disciplines come to understand themselves. And let us suppose that philosophy, when entering such a discussion, has the additional task of defining the manner in which such change occurs. How, then, will philosophy have to redefine itself in terms of the desired third possibility of interacting with other disciplines and thus justify the symposia we enjoy?

These meta-level questions will determine my approach to a historical discussion of medical, psychological, and legal theories which have long since become outdated. I shall remain strictly within the terminology used by the 19th century writers, and I shall confine my attention to only a few aspects of the enormously complex problem of medico–juridical relations, restricting myself to the criminal law and examining chiefly the medical view on the issues it raises. Yet when discussing questions concerning the legal responsibility for criminal acts, insanity, and evil intent, I shall try to imagine how a philosophical position which neither 'totalizes' (i.e., follows the first alternative quoted above) nor 'delimits' itself (thus pursuing the second), once engaged in an encounter with other departments of knowledge, could realize the supposed sense of that encounter by bringing itself into question. I shall investigate how critical analysis of philosophy's own assumptions as a discipline may disclose its bases within (or may render it transparent to) the other disciplines it encounters. As a consequence, what were once taken as the primary elements of such a philosophy could then appear beyond its comprehension. And if this philosophy would thereby find itself deprived of its fictitious autocracy, then the initial supposition that interdisciplinary confrontation makes sense could be specified as positing an opportunity of reappropriating lost foundations. A reflection could be inaugurated which neither dwells on what seems outside nor circulates around what seems inside philosophy, but one which reflects philosophy in the mirror of its neighboring disciplines. It would be the silent dialogue of such reflection which could justify a spoken dialogue to render it explicit.

This reflection will be analyzed in three layers of increasing sophistication, which will be developed by discussing three basic aspects of the medicolegal discussion: I. Responsibility for Criminal Acts, II. Responsibility for Insanity, and III. Responsibility for Evil Intent.

I. RESPONSIBILITY FOR CRIMINAL ACTS

The medicolegal discussion of insanity is centered on the definition of that mental state which would preclude criminal imputation for an offence. Assuming – as our hypothesis – the value of a philosophical consideration of this discussion, which issues might be likely to solicit attention?

First Issue: (Thesis:) In Anglo-American law, the presence of reasoning power ([13], p. 423; 39, Ch. 27), understood as the capacity to distinguish right from wrong, defined the condition for responsibility. Only a total loss of that power warranted acquittal, and pleas of lunacy were considered only if the deed was not committed during a 'lucid interval' ([24], pp. 39, 33). The determination of such intervals was then based on sound behavior, consistently observed in the patient. (Antithesis:) Against this practice it was argued that patients were usually not observed long enough in order to establish conclusively the existence of a 'lucid interval,' and that one should not confound sensible actions and lucidity since 'an action might be sensible in appearance without the author of it being sensible in fact' [29].

It might appear (through philosophical reflection on this argument), as if the original inference (in the thesis) from manifest uses of the reasoning power to its universal presence in the defendant's mind involved an inference from isolated actualizations to the existence of an enduring capacity. The insistence (in the antithesis) on the absolute difference between 'acts' and 'states' might be taken to reveal a notion of mind different from that assumed in the original position: For it seems to imply a discontinuity between the criminal acts, now understood as appearance, and the internal states, now understood as reality. However, these philosophically interesting implications of the antithesis, rather than superceding those suggested by the original understanding, depend on its very presuppositions: It was argued (for the antithesis) that a crime committed during a supposed lucid interval should not be punished 'as if the delinquent had no deficiency at all' ([5], IV, p. 25), but rather taken as evidence against the inferred temporary cessation of madness. But, for this evidence to outweigh the general presumption of sanity underlying the legal system ([50], vol. 3, p. 555), one had to infer again (as in the thesis) from past manifestations of disease to its enduring presence,[2] which was manifest again at the moment of the offence. Hence, both arguments presume the same general link between actions and mental states, and the 'philosophy of mind' is not an issue between them.

Second Issue: (Thesis:) Before the M'Naughton trial, the English law assumed that a knowledge of right and wrong in general is sufficient to establish the criminal character of a deed. (Antithesis:) Yet the Scottish law[3] required that this knowledge refer to the very act in question, and in this spirit Lord Erskine argued that in English adjudication as well delusions should be considered as grounds for a plea of insanity ([51], p. 8f.).

It might appear (in a philosophical reflection on this argument) that the ascription of guilt had first (thesis) exclusively concerned the deed itself, and that the perpetrator entered into consideration only when punishment for the criminal act was to be either allotted or withheld. When, instead (antithesis), the relation of the perpetrator to his act – or his intention – was required to enter into the very definition of an act as 'criminal,' the moral question seemed to shift towards the agent and his own consciousness. Again, however, this philosophically interesting implication does not become operative in the actual argument. What is available (even in the antithesis) for examination in a strict sense, is only the deluded beliefs of the defendant when questioned. In order for these beliefs to preclude an evil intention (*mens rea*) and in order for the seemingly rational premeditation to count as 'insane cunning,' it was necessary (in the majority of cases) to assume an unsound mode of thought caused by delusions[4] and thus to argue (as in the thesis) on the assumption of a general state of disease.[5] But then again the acquittal is demanded not because a particular act can be shown to lack the characteristic necessary to *make* it 'criminal' (i.e., an evil intention), but is demanded only on the grounds of general insanity *excusing* a criminal act. Hence, the philosophically relevant regress to intentions was merely a polemical device for bringing into question the original presumption that partial insanity (or the existence of a general knowledge of right and wrong in conjunction with a mistaken belief about the act in question) cannot establish a successful defense of insanity.

Third Issue: (Thesis:) The criteria of 'reasoning power' in the English law or of 'voluntariness' in the French *Code Pénal* presuppose a unity of mind which is meant to constitute its rationality. A knowledge of right and wrong should imply the power to act accordingly, or the willing of an act should betray an antecedent conscious choice. (Antithesis:) Against this assumption Pinel and later Prichard[6] argued that a partial impairment may affect only some isolated faculties of the mind, while the others continue their usual function. In moral insanity and its subdivisions, the will alone is driven by a particular impulse, which disrupts the rule of reason. Any imputation of a motive, as it implies an antecedent sanction of the act by rational self-regard, is thereby excluded.

It might appear as if the unity of the mind, as taken for granted in the legal understanding (thesis), must yield to a more detailed scrutiny (antithesis). Only the cooperation of various independent faculties insures the consistency of the whole. (One could, then, speculate on the consequences of such a view for the renowned problem of personal identity.) However, if viewed clearly, the (antithesis) alternative suggestion in turn remains dependent on the assumption of that very solid substantiality (thesis) which it had contested. What is (even for the antithesis) accessible to a legal examination in a strict sense is usually a variety of possible motives adduced by the prosecution.[7] In order to prove that such motives – even if they had occurred to the defendant's reason – had not induced his will to cooperate, or to argue for partial insanity, one had to establish independent evidence concerning the accused's supposed derangement. Thus (just as in the thesis) another substantial unity – though now of an afflicted mind – is taken to establish the disintegration of its various faculties. Hence, both arguments proceed on the same kind of inference from the (healthy or diseased) state of the whole to the valuation of the function (or dysfunction) of the parts.

These three issues chosen from the complex net of arguments in the medico-legal discussion have focused on three basic questions concerning the existence of: (1) a capacity to be rational, (2) and evil intention, and (3) the operative force of a motive. In none of the cases did the philosophical view of the argument succeed in illuminating its actual solution. Each Antithesis, while seemingly replacing some uncritical assumption in the original Thesis by greater philosophical sophistication, was successfully defended as applicable in court only when in turn endorsing those same uncritical assumptions. Thus the reasoning which might have justified such a defense was not really operative in rendering it successful, and philosophy's hypothetical interest in the medicolegal discussion was misplaced.

Against that result, however, one could argue that it is the very inability of each of the antitheses considered above to present a genuine alternative to each of the theses criticized, while not illuminated *by*, is yet illuminating *for* philosophy. This inability may then be taken to reveal a peculiar conceptual structure, reminiscent of the manner in which cognition may be philosophically understood. For philosophy, objects of investigation seem *given immediately* as ideas, or as perceptions, of which the mind is intimately conscious. But these ideas are meant to count as objects of knowledge only insofar as they *point* to something beyond themselves, more real, more stable, yet not immediately accessible to the perceiving mind. This understanding of cognition places the 'cognizing' subject 'in between' what seems *available* within and what seems *signified* without – as the mirrored image, while immediately present on the surface of the looking glass, *counts* only as reflection of the spatial thing external before it, but behind the one who is looking from a position 'between' the image and the object. (The notions underlined in this paragraph henceforth will serve as technical terms and they will be used to remind the reader of the conceptual pattern introduced here.)

In the context of the medicolegal discussion as it concerns the *knowability* of reasoning power, evil intention, and motives in another's mind, this internal-external relation might be seen to imply the need to argue indirectly or from contextual assumptions. Thus (for the first issue), the legal proponents of the thesis had to be reminded that the observable (what was *available*) for or against a (*signified*) lucid interval *points* toward a state not open to inspection. In a reported (*available*) delusion (i.e. the second issue), just as a senseless crime (i.e. the third issue), cannot *immediately* establish the lack of motive (third issue) or bad intention (second issue) unless each is accepted as a symptom of the unobservable (*signified*) disease ([28], p. 62). Such reference to the context of interpretation, since it had been disregarded in the original legal definitions (thesis) could only then be effectively confronted with an alternative context of interpretation (endorsed by the predominantly medical criticisms or antitheses quoted above) when its necessity was rendered explicit. As long as medical experts had accepted the challenge to model their evidence on the legal principle 'de non apparentibus et non existentibus' ([39], p. 401), that is on a principle that denied the necessity of indirect reasoning, they failed to make their point ([4], p. 584f.).

But even when insisting on that necessity, a realm of medically accepted, available data had to be defined. It became imperative to develop uniform categories for a consistent classification of mental illness based on clearly distinguished characteristics.[8] Only given such *immediately* perceivable (though in most cases scientific) evidence could one defend their merely *signifying* function as symptom-complexes, where each element taken for itself conveyed no information, but where their interrelation constituted that quasi-organic unity which the experienced practitioner is accustomed to interpret. If (first issue) insanity is experimentally proven to be a physical disease[9] and if pathological states must therefore be supposed to prevail even while mental symptoms are absent,[10] a confirmed lunatic must be considered *non compos* (or in a state of heightened irritability) even during what appears to be a lucid interval.[11] Equally (and this argument constitutes a consequence from the second antithesis), since any affliction of a portion must (on the basis of scientific evidence) sooner or later infect the neighboring regions of the nevous tissue, confirmed delusions, even if seemingly not connected with the offense, must point to its physical causes.[12] Similarly (and this argument is meant to prove the relevance of the third antithesis), the apparent soundness of intellectual faculties during confirmed partial insanity will sooner or later be diminished, their organs being infected by the diseased moral organs, and thus reveal the pathological nature of the violation.[13]

The confrontation with the law and the necessity for medical experts to convince a jury ignorant of the state of science ([28], pp. iv, 4, 6f.) contributed to the development of a specifically psychiatric terminology and pattern of diagnosis by which to effectively defend that alternative context of interpretation which was to replace the original one. The concept of insanity as it was originally understood – as *dementia* and loss of reasoning power – had rested on an implicit agreement between the medical and the legal view.[14] But as brain-dissections and a closer study of asylum inmates revealed that more are sick than just the raving, an ever-increasing variety of mental diseases had to be admitted into medical knowledge. In contrast to the legal presumption of the rational unity of non-demented minds ([28], p. 17f.), the dissolution of the mental life into discontinuous series of events (first issue) and of the brain into a multiplicity of separate organs (second and third issue)[15] marked the beginning of a dis-association of the medical concept from the legal concept of insanity.[16] This dis-association was completed only when an alternative unity of mind[17] was re-established, in purely physiological terms.

Returning to a meta-level consideration (cf. above, p. 16), but now from the standpoint of medicine, this professionalization of psychiatric knowledge made alternative interpretations of the relation between medicine and jurisprudence possible and thus accounts for another layer in the medicolegal dispute. As a first possibility (reminiscent of philosophy totalizing itself), one argued thus: Strictly speaking, no clear distinction between 'sane' and 'insane' can be medically drawn. Just as the term 'disease' in general applies to an infinite continuum of subtle shades ([11], p. 216ff), so nobody can with certainty be excluded from a suspicion of some limiting degree of irritability and thus of potential mental illness. In this vein, a priority was claimed for the medical concept of insanity over its more narrow application in court, and an attempt was made to reduce what is implied in the latter to what is known in the former.[18] Friedreich even held that the medical expert should not simply state pathological facts but should decide concerning the freedom and responsibility of the defendant ([19], pp. 16, 54). Thus, he arrogated to medicine the competence of a judge and made jurisprudence a branch of universalized medical knowledge. On the other hand (as a second possibility, reminiscent of philosophy delimiting itself), the practice of examining the sanity of a culprit only after the court had declared his guilt and of allowing the Home Secretary to overrule the verdict of the jury totally divorced the medical investigation from the judicial concerns. The expert restricts himself to purely pathological facts, expressed in medical terminology and incomprehensible to the general public ([7], [25], p. 238). The discipline maintained its integrity, but it pronounced judgments which affected the legal system without being controlled by it.

Thus, within medicine's self-understanding a duplicity similar to that found in the philosophical views initially quoted can be discovered. But if, in the case of philosophy, an adequate compromise must yet be developed, the relation of medicine and jurisprudence which finally prevailed may serve as a paradigm. In the course of their discussion the intermediate stage polarization of the respective medical and legal definitions of insanity yielded to the acknowledgment of a double meaning of the term: 'insanity' as a gradual scale of universal application was distinguished from 'insanity' as a criterion signifying that lack of the practical freedom presumed by law.[19] The requirements of the courts were served while the expert could remain on the ground of his competence.

We find, then, on the metalevel of our discussion and at the intersection of medicine and jurisprudence, a conceptual structure disclosing itself which again mirrors the philosophical understanding of cognition (p. 20). In a strict sense, only insanity-by-degrees is given (*available*) to medical knowledge and is accessible to its categories. But this is seen by medicine itself to acquire a *significance* beyond the limits of the discipline. Insanity-as-criterion, since it causes acquittal from punishment and excludes responsibility – the very concepts of which are beyond what is available within the medical horizon – must be proven (as *signified*) by medical evidence plausible in the eyes of the public. This double conception of insanity allowed medicine to view itself not only as a discipline in its own right, but also in a larger social context where its own concerns had a novel application – in short, to understand itself as an interdisciplinary endeavor.

As a first result of our philosophical appraisal of the medicolegal discussion, it appears that (on the object level of that discussion) one could justify not only the inability of the antithesis to present not merely a substantial alternative context of interpretation, but also its failure to present a structurally different approach to minds. One could also (on a meta-level of that discussion) account for the ability of the medical profession to view itself 'in proportion' with jurisprudence. Medicine here discloses its own bases to the competence of legal science, thus becoming 'transparent to' or reflecting itself in the notions of the law in a sense of 'reflecting in' which our title suggests. Hence a paradigm has been established for the task we have set of imagining an understanding of philosophy appropriate to an interdisciplinary dialogue.

II. RESPONSIBILITY FOR INSANITY

Mental illness established an act as non-punishable insofar as it precluded moral choice. But the question remains open whether the occurrence of the illness in the first place was the responsibility of the defendant and how such responsibility should affect the evaluation of subsequent offences. If, for a long period of time previous to the medicolegal discussion, the few available definitions for mental illness seemed sufficient, it was largely because a general consent prevailed that whoever was not totally bereft of mental powers had impaired his mind by some vicious act or habit and thus – in the event of a crime – was liable to punishment.[20]

During the 19th century such reasoning was slowly abandoned. For the philosophical consideration of responsibility, however, it might still appear to be relevant. As with the interpretation of offences during insanity in general, so in the search for the cause of insanity, one might ask whether the ability to reason implies an additional decision to reason in a given case and whether this decision depends on a further capability to apply reason, etc.[21] Thus, for example, one might ask whether it was Nicholai's superior moral character (or a moral decision) that prevented him from yielding to his delusions or whether the general strength of his faculties (or a natural capability) was the cause of his sensible skepticism.[22]

What appears to be the medical answer to this question, however, repudiates the philosophical statement of it. The argumentation of experts in court implies that deviant mental behavior is caused by physical irritation, and that such irritation explains the pathological nature of such behavior. Correspondingly, a consistently sensible mind betrays a normally functioning brain. Statements on mental phenomena can be considered as imprecise descriptions of what are strictly speaking only physiological facts about the brain. From thoughts and feelings as mere appearances, the real matters must be reconstructed. Such physiological determinism[23] implies that mental illness cannot be induced by a wrong morale. But this seemingly ontological commitment is defended by merely practical considerations ([28], p. 78; [10], p. 76): One cannot reason a monomanic out of his errors ([47], p. 187), but purgatives, tepid baths and a balanced regime work miracles on his 'metaphysical' convictions. Hence the philosophical puzzle over priorities between natural capability and moral decision conjured above is no issue.

Yet, it is also argued ([11], p. 142ff.; [9], pp. 173, 177; [4], pp. 26f., 29f., 239ff.) that social unrest, wars, revolutions, religious fanaticism, novel ideas, and the loosening of familial bonds, by producing more violent passions, increase the danger of mental disease. For its prevention, a sedate, orderly, virtuous life and balanced exercise of all the mental powers was recommended. At least some of these circumstances rely on an understanding of meaning, prior to their physiological effect upon the brain. Furthermore, the experience of the medical practitioner in dealing with some types of mental patients testifies to the beneficial influence of a 'moral management.'[24] This mode of treatment extended not only to control by punishment and reward, but to gaining the patient's confidence and respecting his sense of honor, to kindness and sympathetic understanding as well.[25] Even if such advice could be adequately reduced to rules for conditioning the environment, some intelligible signals have been admitted to be of consequence for the state of the brain. Thus, within the medical position a contradiction seems to exist between a manner of arguing grounded on exclusively physiological matters and a manner of acting on the tacit assumption that mental events do make a difference in their own right.

When seen in the context of the medicolegal discussion, the question of responsibility for insanity leads to two quandaries: *First*, two connected questions were answered in opposite ways by the medical and legal professions. *Second*, medical practice in dealing with mental patients implicitly contradicts the kind of theorizing which underlies the answers referred to above.

First Quandary: The original legal presumption that a person is responsible for becoming insane implies that the question (A): 'Can a mind be blamed for what happens to its brain?' is answered 'Yes.' This answer implies in turn that the question (B): 'Is the mind reducible to the brain?' be answered 'No.' The medical criticism of that original presumption rests on the fact that question B is answered 'Yes,' which implies that question A is answered 'No.'

Second Quandary: Yet the medical practice of *treating* insanity by also *appealing* to the mind implicitly contradicts the medical 'yes' to question B, and the medical practice of using 'moral management' for such treatment implicitly contradicts the medical 'no' to question A.

A philosophical view of these quandaries may not be called for on the basis of the respective positions themselves. Even the inconsistency generating the first quandary can be excused by the practical need for treatment in spite of the undeveloped state of physiology. But such a view is clearly required on a meta-level and on behalf of philosophy itself: Our hypothesis that philosophy might wish to engage in a confrontation with medicine would hardly be tenable if the medical reduction of mind to brain were left unanswered. This reduction must appear as a threat not only to man's dignity as a moral agent (and thus to the subject matter of philosophical ethics) but also to the very integrity of the philosophical discipline itself.[26] If philosophy rules over ideas and if ideas in the tradition of modern philosophy are considered as mental, then the abolition of mind deprives philosophy of its authority and exposes it to the kind of ridicule directed at a king without subjects. While a philosophical view of the problem of responsibility for insanity had to be admitted to be irrelevant to the outcome of the argument, a philosophical view of the argument's structural complexity may yet be helpful to avert that meta-level threat. If by interpreting that complexity in the light of the cognitive relation, the insolubility of the conflict (first quandary) can be accounted for, and if the contradiction within the medical position (second quandary) can be removed by modifying that position, then a more reflective understanding of philosophy's subject matter and of its reducibility to medicine becomes conceivable.

First Quandary: The conflict between the medical and legal answers to questions A and B is insoluble, because both questions are, conceptually speaking, insignificant. The reducibility of mind to brain is no issue, because both concepts, when seen in terms of their cognitive relation (p. 27), denote abstractions. The reductionist thesis does not deny that we believe we have a mind or that we are conscious of something mental. It does deny, however, that the 'real referent' of that belief is some real thing, 'Mind'. Yet, since even thinking about 'a mind in itself', or 'had' by none as the real referent of our consciousness is inconsistent, not much has been denied. The in-between-ness which had characterized the subject of knowledge precluded its object being posited either totally within or totally without the grasp of comprehension. Mind, on the other hand, is by definition entirely inside, immediately accessible to its own awareness. Whereas we can have ideas *of* the brain suggesting the brain as the real referent, we can have ideas only *in* mind. Moreover, while we may be *conscious* of mind as such, all *knowledge* takes place in the interspace between what is given and what is suggested. Thus the mind, as it is immediately *given*, does not qualify for its own knowing consideration unless it suggests (*signifies*) to itself what is principally outside of and inaccessible to it (and thus provides its quasi content), but which at the same time is believed to constitute its stable reality – namely, the brain. The brain, because it provides such a stable condition for thought and the real instrument for its application, may be taken to occupy that vacant space 'behind' that cognizing mind which, in following the suggestion of its self-image, becomes transparent to itself. Yet brain is here referred to not as an object *of* which, but only as the instrument *by* which we can have ideas in mind. It follows only that mind, as independently *given* is *knowable* only as it signifies brain. It does not follow (as in the reductionist thesis) that the only givens qualifying for knowledge are our ideas of the brain as signifying brain. Hence, a negative answer to question B either denies what is not even worth supposing, or it denies too much.

The same cognitive interpretation exposes question A as quasi ungrammatical. This imposes questions concerning the knowability of actions (like 'damaging' or 'neglecting') of one entity (the mind), upon another (the brain), while neither is even conceivable as an 'entity' without implying the other.

Second Quandary: However, in the context of the inconsistency of the medical position, this question (A) arises again. On the one hand, mental patients are treated as if to blame, on the other, as if not to blame. The question under what conditions one should – practically – blame refers back to the question under what conditions one can – theoretically – attribute blameability in a sense in which one might attribute freedom or responsibility.

None of these properties makes sense when attributed to either minds or brains, but for different reasons. They are not attributable to brains, because these are objects of scientific study only insofar as their states are conditioned. They are attributable to minds, because minds, when cognitively interpreted, are not the kind of things to which one attributes anything. Freedom and responsibility are *given immediately* in a sense in which the mental itself is given. Hence they can only be appealed to, or challenged, by the recommendation of norms. Such norms underlie the legal position as legal principles and the medical position as the standards of good health. Yet such an appeal can be *known* to be effective only when they – again – *point* beyond themselves, or *signify* some conditioning treatment as their real reference. Thus the law, even though it professes only to be presenting normative appeals, enforces them by the persuasive power of punishment. And thus medicine, even though it professes only to treat diseases, can quite consistently be conceived to render effective thereby its own appeal to norms. As a consequence, even if the moral management of the insane can be reduced to a pragmatic strategy of conditioning (what appears to be) a mind for the treatment of the brain, still the physical treatment of body and brain can quite consistently be conceived as a strategy of persuasively enforcing an appeal directed to a mind. But, then, even that pragmatic strategy of treating the brain by moral blame may in turn be conceived as persuasive means of challenging the mind.

On the meta-level of our considerations, the mind–brain issue demarcates a space of intersection between philosophy and medicine. The seeming contradiction between medical theory and the implications of its practice points only to a corresponding complexity in medicine's own foundations. Strictly speaking, only the brain is accessible to medical knowledge. But the brain enters into such knowledge only as conceived by an investigating (medical) mind. Thus, while it is not inconsistent for a physiological reduction of the mind to the brain to be entertained by a mind, it does expose the conditions under which medicine views its objects to a cognitive interpretation and thus to the judgment of philosophy. As a second result of our philosophical inspection of the medicolegal discussion, it appears that what was presumed to be medical fact lacks that absolute status within medical knowledge which would qualify it to repudiate philosophy. It also appears, however, that philosophy in itself lacks that conceptual integrity which might have made it worth repudiating. For if the mental is knowable only as transparent to the brain, and if the brain belongs to the competence of medicine, then philosophy's rule over mind must be shared.[27] Its self-sufficiency as a discipline for ideas is exposed as fictitious, and its interdisciplinary nature can be acknowledged. Thus, in its relation to medicine, philosophy follows the paradigm which has been set by medicine in relation to jurisprudence. While reflexively applying the cognitive relation to its cognition of the mind, philosophy comes to view its 'own' concepts as legacies from medical knowledge.

III. RESPONSIBILITY FOR EVIL INTENT

Even though the meaning of freedom has not yet been specified philosophically and even though it can be specified medically that freedom is knowable only as the functioning of a healthy brain – in the eyes of the law a choice is considered free and responsible ([43], p. 190) if it is rational, i.e., made in accordance with reason and the 'sentiments' ([20], p. 67). Reason proclaims the right or wrong of actions to which the 'sentiments' incline. When contemplating an illegal action, it subtracts the pleasure of that action from an expected lawful punishment to determine the remainder of pain. A crime committed implies that the choice was wrong; its punishment can be seen as a persuasive argument against poor calculation, suggesting to others the necessity of keeping more accurate accounts.

From a medical view, however, such incorrect estimation might always have mental impairment[28] as a cause. Supposing, in light of the legal argument for deterrence, that no incurred advantage can outweigh capital punishment, and supposing the relevant facts and their legal implications are known, then the criminal act must either be due to a lack in the calculating power, which excludes the required rationality, or to an irresistible impulse of desire, which suggests a diseased state.

Yet it was this very assumption which exposed the medical profession to the charge that it annihilated the basis of criminal justice ([21], p. 29; [13], p. 364; [49], p. 34) – a reproach which gave the legal writers a powerful argument against accepting the medical view of insanity in court.

Arguing against this reduction of crime to derangement, the legal authorities insisted upon the possibility of malice in a healthy mind. Though the law does not deal explicitly with morality, the imputation of a crime does imply the supposition of some wicked purpose.[29] Thus, a distinction between murder and manslaughter is maintained, and external force, intoxication, and even violent passions are pertinent to the determination of guilt. The punishment is proportioned to the amount of evil intent necessary to explain the act, such malice being understood as the choice of a private good, despite its ill effects on the public.

The problem, then, of responsibility and evil intent arises because the question 'whether crime and mental health are compatible,' is answered in opposite ways within the legal and the medical positions. That problem raises the philosophically interesting question concerning the meaning of rationality, and its relation to malice and disease. Moreover, to a philosophical mind the conflict between both positions is insoluble, for both rest on inconsistent arguments. Here, just as in the first two parts of our philosophical reflections on the medicolegal discussion, the actual outcome of the argument (the allocation of 'serious' cases to disease and of lighter cases to immorality) does not itself depend on any philosophical reasons. But the inconsistencies within each position in the former two cases could only be philosophically *interpreted* in terms of the cognitive relation, and the relevance of that interpretation remains conditional upon the relevance of the meta-level (i.e., conclusions concerning the manner in which medicine and jurisprudence (Part I) or medicine and philosophy (Part II) relate to one another). However, those inconsistencies derive from inadequacies within philosophy's very own understanding of the idea and the reflective nature of rationality. Hence the analysis of these inconsistencies naturally falls under the competence of philosophy and will reflect back on that discipline understanding itself.

First Quandary: The inconsistency within the legal position consists in the contradiction between the practical opposition between the legal and the medical means of 'appealing to responsibility' (punishment versus treatment), and an incapacity to explain theoretically the grounds of that opposition. This incapacity arises, because calculation of pleasures and pains is an inadequate basis for knowingly assessing a person's responsibility.

In the argument for deterrence, the sanity of a mind is thought to reside in its power to calculate pleasure and pain. Only a false calculation could motivate the choice of selfish pleasures over a well-enforced public good. Disregarding any impairment of the mental faculties and allowing in usual cases sufficient time to contemplate the advantages to be gained on each side, the final choice can only be explained by a particularly evil intent. But it is exactly this notion of healthy evil which remains incomprehensible. In cases where all selfish regard is lacking, where no guiding interest can be traced, where the agent does not enjoy his atrocity, where evil is sought not for another good but for evil's sake alone, the model of balancing pleasures and pains breaks down, and one is reduced to assume with the medical writers some natural depravity which cannot be blamed.[30]

This inability within legal thought to account for rational malice reveals a lack in the presuppositions for legal categories, for which juridical science is not to blame. The concept of rationality belongs to the proper domain of philosophy. But in the tradition of philosophy, the ideas of rationality (or truth) and of the will were never in principle separated from the idea of the good. Nobody is voluntarily bad ([41], *Timaeus*, 86c) and what is the object of desire is thereby defined as good. This good, furthermore, can always be placed in a hierarchy where each particular good is better or worse in relation to others and where to the highest good only a lowest good corresponds – not an evil. Badness is the choice of the lesser good,[31] and is reducible to error, but 'ignorance is the greatest of diseases' ([41], *Timaeus*, 88b).

Thus, in the mirror of the legal concept of rationality, another mirror of philosophical definitions appears to which the legal notion points, as if to another object. One of the presupposed notions within jurisprudence is thereby rendered transparent to one of the thematized notions of philosophy, and the inadequacy of the former suggests an inadequacy in the latter. Philosophical assumptions enter into the terms by which other disciplines appropriate to their concerns ideas they cannot themselves establish. If these terms, when applied to such concerns, fail to do them justice, some claim of philosophy's competence over its own concepts has been repudiated. It becomes, then, incumbent on philosophy to disassociate what seemed its proper subject from its pretended comprehension or to render it transparent to yet another discipline, – such as, for example, religion.[32] In religious thought, the 'good' may be taken for granted and does not get identified with what is rationally comprehensible in such a way that what is utterly 'not good' must as well escape such comprehension.

Second Quandary: The inconsistency within the medical position arises not from an inadequate idea of rationality (as with the law), but from an inadequate understanding of the reflective nature of appeals to rationality. The diagnostic situation in psychiatry is mistaken by medical experts themselves as requiring the ascription of rationality or irrationality to other minds. But as the distinction between mind and brain in the second part of our reflections suggests, what can be *ascribed* is only health or disease to a brain. While in the law it is by means of a threat of punishment, in psychiatry it is by means of such ascriptions that one may then *appeal* to the rationality of a mind.

In both cases, such appeals rest on the ability to criticize judgments, either concerning the adequate estimation of pleasures and pains incident to an action, or concerning that view of the world which must be presumed responsible for that change of character without adequate external cause[33] universally quoted as a decisive criterion for insanity. Consequently, these appeals can be understood as endorsing the in-between-ness of the rationally cognizing subject[34] which minds the balance of what it knows to be known and what it knows to be not known.[35] If ideas only point, or suggest, but do not capture their real reference, then they are correctable in view of the judgment of others.[36] But whereas in the law such correction is effected by an act of punishment, in medicine it rests on holding up a view of how the world really is and on treating the patient until he has accepted that view. It is thus supposed that the physician can place a fool and his beliefs about the world in relation to the world, and that the appropriate measures follow from the difference between the two. But, philosophically speaking, he has placed only his view of the fool in relation to his view of the world; what is inside the judge's mind is meant as an outside criterion by which to examine the mind of others. Hence 'insane' is not a property attaching to the one so defined, but is equally a mirror in which social judgments are reflected. (Foucault [15] has shown in detail how to judge a society in terms of its judgments on insanity.)

Moreover, while the judge supposes himself to be rational because criticizable as to his world view, such criticizability (or such condition for knowledge) does not enter into his appeal. Unlike in law, in medicine the one to be corrected is denied the very competence to correct back (as Thomas Szasz has shown). Hence the medical expert, when applying his – philosophically reflected – standard of rationality while identifying that standard with his own mind, himself becomes a prey to folly.

Yet this inconsistency, when exposed by a philosophical reflection holding up a mirror in back of the medical mind, equally reflects back on philosophy. After all, it is the philosophical habit of arrogating to one discipline the right to legislate rationality that is mirrored in medical practice. If it is then a philosophical inconsistency in understanding its own rationality which underlies the inconsistent medical position, then it becomes incumbent on philosophy to render its own world view criticizable, or to render its own subjects transparent to the knowledge of other disciplines. The mirrored image reveals yet another mirror at the opposite wall, and, while contemplating infinite reflexivities among other disciplines, philosophy finds itself exposed from behind and thrown into that picture from which it had seemed to maintain a theoretical distance.

As a result, the interdisciplinary confrontation holds up each discipline as a looking glass by which to expose the back-side of each other's minds. The need arises, in order to seek cover, to secure one's mind's 'behind' by rendering it transparent – like a mind to its brain – to the other disciplines. Thus, philosophy, as the master and minister of rationality, may now itself be questioned in terms set by those disciplines. And as medicine might have profited by questioning its own criteria for unsoundness of mind in the light of philosophical rationality,[37] so this rationality in turn will profit by questioning its medical soundness.[38]

Indeed there seems to be ample evidence for the relevance of such a questioning[39]: has not philosophy occupied the mind for centuries with one single and inconceivable idea of 'truth'? Do not philosphers have souls that continually talk to themselves? Do they not strive for what is impossible? Do they not suffer from cold extremities, headaches, and poor digestion? Do they not hallucinate when seeing atoms dancing in the void and idols flowing from things? And what about those delusions of substantial forms, things-in-themselves, Beings, and Nothingness?

But there is hope. It was universally accepted among the medical writers of the time that nobody can be a fool with respect to that which he considers his folly. And even though philosophers, unlike the more domestic fools, knowingly insist on their rational dreams, they do so in order to keep that reflection alive which presents a structural condition for soundness. Thus, our initial supposition that interdisciplinary confrontation makes sense if the desired third possibility of interacting with other disciplines can be specified, has been confirmed. As rationality has been interpreted (in view of the cognitive relation) as the realization of one's own position 'in-between' what is accessible to comprehension and what is signified behind such available data – so, on the level of academic fields, such rationality consists in avoiding both extremes of hermetic self-containment and of an uncritical extension of competence, and in placing a discipline in between what seems to define its very own subject matters and their reference to the knowledge of other disciplines behind.

ACKNOWLEDGEMENTS

I would like to thank Henry W. Johnstone, Betsy Garlitz, and Douglas Booth for their generous criticism, as well as W. T. Cooper and H. Tristram Engelhardt for their reasonable and sensible linguistic suggestions respectively. John Moskop's final corrections and Chester Burns' initial help in collecting bibliographic information were very useful. All the mistakes I have introduced later are my own business.

The Pennsylvania State University,
University Park, Pennsylvania

NOTES

[1] L'objet de cet ouvrage est surtout de . . . préserver la médécine du mal que peut lui faire une secte philosophique . . . ' ([6], p. viii); see also Georget ([20], p. 67), Ray ([43], p.55), and Beck ([4], p. 614).

[2] Haslam ([28], p. 13ff.) discusses the difficulties involved with the attempt to establish a reasoning potential.
[3] Beck and Beck ([4], p. 604) also quote Hume and Alison for this interpretation.
[4] See Becker ([50], p. 463ff.) and Hamilton ([25], p. 234). Wharton and Stillé ([49], p. 50) emphasize the need to 'take a *sane* view of the right and wrong' (italics added). Fisher ([50], vol. 3, p. 147ff.) quotes the case of Wilhelmina Lebkuchner, who poisoned two of her children in order to save them from a life of misery such as she herself had experienced. When considering her own misfortunes, one cannot help agreeing that she had 'good reasons' to believe death to be a preferable state. But on a closer inspection it appears that it was her quite unreasonable beliefs concerning the West Coast, where her children were to be sent, which ultimately motivated her act and which reveal what might have soundly been considered signs of an unsound mind.
[5] The same argument for the evaluation of motives in a defendant's mind can be found in Haslam ([28], p. 14).
[6] See Prichard ([42], p. 21), Ordronaux ([39], p. 426), and Georget ([22], p. 35): '... des folies sans délire, des lésions exclusives des penchans et des sentimens ou de la volonté, qui provoquent à des actes insensés ou atroces que la raison reprouve ....'
[7] On motives in the insane mind, see Haslam ([28], p. 53), Esquirol ([14], p. 356), Ray ([43], p. 45), and Wharton and Stillé ([49], p. 115ff.).
[8] On the need of a clearer nosology, see Spurzheim ([47], p. 3). For the abuse of medical expertise due to the prevailing variety of explanatory hypotheses for any given case, see Wharton and Stillé ([49], p. xiii).
[9] Physical criteria quoted by Haslam ([28], p. 49) are, for example: former attacks of insanity, family history, injuries of the head, mercurial preparations, attacks of paralysis, suppression of customary evacuations.
[10] Friedreich ([19], p. 414) discusses the concept of 'disposition'.
[11] Georget ([23], pp. 54, 94), Prichard ([42], p. 272f.), Beck and Beck ([4], p. 596f.), Friedreich ([17], p. 61), Ordronaux ([39], p. 401) are only some to hold this opinion. For the term 'irritability' see Broussais [6].
[12] See Combe ([10], p. 268) and Chapman ([8], chap. xiii). Against this Elwell ([13], p. 396) holds: 'A man whose mind squints, unless impelled to crime by this very mental obliquity, is as much amenable to punishment as one whose eye squints.'
[13] See Esquirol ([14], p. 334) and Combe ([10], p. 272f.). Hamilton ([25], p. 230) argues from the existence of *any* insanity to the possibility of violent acts. Thus to establish the former is tantamount to excusing the latter.
[14] Hale ([24], p. 29ff.), when characterizing the *legal* understanding of *dementia*, refers to *pathological* facts. Similarly, the term 'unsound mind' was coined by legal writers to signify an incapability to manage one's own affairs, in addition to that mental impairment which was required to establish a *case* for the chancellor to grant a committee for protection of person and estate. No pathological symptoms were conceived to correspond to this *case*. Its only function was to 'match' a quasi-medical language with a fact pertaining to civil behavior which could not be subsumed under 'idiocy' or under 'lunacy.' See Haslam ([28], p. 75ff.). For his criticism of this term see p. 81ff.; for his reduction to *medical* terminology, p. 89.
[15] Combe ([10], p. 4) follows Gall when he states: '... every organ in the animal economy performs a separate and appropriate function ....'
[16] Only when insanity – for the purposes of jurisdiction – is no longer considered as a question of law (that is, adequately definable in legal terms) but a question of fact, to be established on medical grounds, is this distinction acknowledged. So Elwell [13] insists on a distinction between insanity as a 'hypothetical form of bodily disease' (p. 397) and what legally excludes punishability. See also Becker and Boston (in [50], p. 509). An unfavorable view of this separation is exemplified by Hammond ([27],

p. vi): 'The law establishes an arbitrary and unscientific line, and declares that every act performed on one side of the line is the act of a sane mind, while all the acts done on the other side result from an insane mind .... But every physician knows that it (the right–wrong test) is not a medical line.'

[17] See Ordronaux ([39], p. 447), who speaks of the 'great sympathetic system' where disturbances in one region will have unforeseeable effects on all the rest.

[18] See Spitzka ([46], p. 23): '... that cannot be sanity in law which is insanity in science.'

[19] Becker and Boston (in [50], p. 437) state that for the law, no degrees of insanity can exist. Already Conolly [11] had – for cases of confinement – established such a distinction between medical and medicolegal insanity (p. 373). As a consequence it became further possible to distinguish moral guilt from criminal culpability; see Georget ([21], p. 7ff.)' '... les fous ont ... conscience d'eux-mêmes, et l'on pourrait aller jusqu'à prétendre qu'un fou peut devenir coupable dans toute la force de cette expression (and without being legally culpable).'

[20] Even the enlightened 'somatist' Battie ([1], p. 57f.) placed some vices among the remoter causes of madness; Conolly ([11], p. 155) includes sensual indulgences; Friedreich ([18], p. 7ff.) quotes Heinroth for the view that all insanity arises out of sin; even Ray ([43], pp. 46, 176) sees in self-caused partial insanity a reason for blame; Elwell ([13], pp. 396 and 410ff.) considers 'moral insanity' and 'depravity' as identical with the vice from which they originated, and holds that even if acknowledged as diseases they can and ought to be controlled by the afflicted persons. Ordronaux ([39], Sect. 189) quotes Wharton, according to whom the 'moral sense has to be created by the state.' However, in the preface to their third edition Wharton and Stillé ([49], p. vii) explain that the definition of insanity had also been kept in very narrow limits in order to reduce as far as possible the number of those who were subsequently subjected to the barbarous treatment of an asylum.

[21] Combe ([10], p. 113) talks of an 'excitement of function inducing disease of the organ,' – and the question remains: who excites? Or in the case of exaggerated religious practices (p. 174): who abuses? Similarly for sanity: who insures the required balance of mental exercise? See also p. 144ff.

[22] See Haslam ([28], p. 25ff.). He quotes a translation of Nicholai's own account: 'I endeavoured to collect myself as much as possible, that I might preserve a clear consciousness of the changes which should take place within myself' (p. 28).

[23] See Broussais ([6], p. 22). 'La sensibilité est donc la consequence de l'irritabilité, tandis que l'irritabilité n'est pas celle de la sensibilité.' (See also p. 62, 86). '... l'idée de liberté ... n'est qu'une formule. Il faut bannir l'entité, et ne voir que les faits: ... il faut de toute nécessité mettre la liberté des malades et celle des fous sur la même ligne que celle de l'homme en santé, car le fou dit aussi *Je suis libre* ... ' (p. 141). Similarly Friedreich ([19], p. 12). 'Die Seele ist demnach nichts an und für sich Selbstandiges; sie wird auf die, dem ganzen Organismus einwohnende, Lebenskraft, wovon Sie ein specieller Ausfluss ist, reduciert.' See also p. 86ff.

[24] See Spurzheim ([47], for example, p. 182): 'No deception ought to be permitted,' also p. 185; Esquirol ([14], pp. 34f., 64ff., 219); Tuke ([48], p. 139): '... that patients should be treated as much as possible as rational beings'; and Hammond ([27], p. 729): 'Great assistance may be obtained through the intelligent co-operation of the affected individual.'

[25] Esquirol ([14], p. 230) endorses a 'médécine morale, qui cherche dans le coeur les premières causes du mal.' For a discussion of the possibility of soul causing its own disease see Friedreich ([18], p. 13ff.).

[26] This connection is seen clearly by Broussais ([6], p. 89) when he reduces those very mental facts, which philosophy is bound to treat as *sui generis*, to physiological matters.

27 Some awareness of this necessary duplicity of views is present in Combe's statement ([10], p. 39) that just as metaphysicians fail when limiting their considerations to the data of consciousness, so anatomists fail when limiting their attention to physiological data, where the close analogies of structure in the various nerves must preclude any derivation of their different (mental) functions. Similarly Spitzka ([46], p. 17f) argues that even though insanity is a disease of the brain, the diagnosis must be 'empirical,' that is, confined to the mental symptoms.

28 See Rush ([45], p. 360ff); Georget ([21], p. 78); Conolly ([11], p. 337ff); Ray ([43], p. 80). Many medical writers indeed admitted that 'men of the highest attainments and most lucid faculties will deliberately commit acts of turpitude' [28], p. 47), but they admitted it only 'because every day furnishes instances' (*loc. cit.*) of what on strictly medical grounds cannot be accounted for. See also Spitzka ([46], p. 43). The frequent disavowal that they would 'shelter crime under the pretense of insanity' is to be seen as a defense against an equally frequent reproach.

29 '. . . a generally wicked, depraved and malignant spirit – a heart regardless of social duty, and deliberately bent on mischief' ([49], p. 806).

30 Georget [20] talks about 'malheureusement nés' and states ([21], p. 126) that an 'acte atroce . . . commis sans motif, sans intérêt, sans passion . . . est évidemment un acte de démence.' In Combe's language it is the same exaggerated size of an organ which constitutes on the one hand a diseased imbalance of the faculties and on the other that precondition which induces to criminal acts. For the legal side Friedreich ([19], pp. 59, 452) quotes Groos (Über Kriminalpsychologie) for the view that the criminal already is not free in the (legal) sense of reasonable self-determination. And (*op. cit.*, p. 62) Sampson (Criminal Jurisprudence) identifies the state of psychic health with the capability to obey laws. Friedreich agrees with both (*op. cit.*, p. 124): '. . . blosse Schadenfreude, grausames Vergnügen, blinder kranker Trieb entspringt nie aus somatisch gesundem Organismus'; (p. 137) 'Es wird sich überhaupt bezweifeln lassen, ob es Menschen gibt, die das Böse nur aus dem einzigen Grunde deswegen tun, weil es böse ist.'; (p. 130) 'Wahrlich die Ehre, die wir der Menschheit schuldig sind, verbietet uns, solche Subjekte als gesund zu betrachten.' As a consequence, as quoted by Georget ([21], p. 21), Grand considers such 'subjects' enraged animals 'que l'on extermine avec raison, pour délivrer la societé des maux inévitables qu'elle souffrirait . . . .' Concerning the (legal) assumption of malice in cases of murder (as distinguished from cases of manslaughter) it is interesting to note that Wharton and Stillé ([49], p. 809f.) find it necessary to distinguish the ordinary meaning of malice as 'spite' or 'hatred,' (which does not concern the law, since that meaning is, indeed, incomprehensible) from its legal meaning as 'a wrongful act, done intentionally, or without just cause or excuse.' They define it as the *negation* of what would be called comprehensible. On the other hand, Fisher (in [50], p. 241) describes the deranged person as one who 'prefers the bad consequences of his act to the restless and unhappy state of mind in normal condition,' that is, exactly as what would usually be considered a criminal.

31 'Die Bösartigkeit der menschlichen Natur ist also nicht sowohl Bösheit . . . als eine Gesinnung . . . das Böse als Böses zur Triebfeder in seine Maxime aufzunehmen, . . . sondern vielmehr Verkehrtheit des Herzens [und diese] . . . entspringt aus der Gebrechlichkeit der menschlichen Natur, zur Befolgung seiner Unbegreiflichkeit' ([32], p. 183).

32 Kant ([33], p. 319), in 'Der Streit der Fakultäten', notes that the concept of evil derives from ideas for which Natural Religion depends on Revelation.

33 The criteria for 'insanity' can be taken to determine the margins of tolerance in a given society, and thereby the limits of sympathy extended to one's fellow men. For instance, Conolly was most generous in tolerating human frailty. Among the medical writers in the 19th century, he stands out in refusing to call Ben Jonson mad on account of his visions – as long as Jonson himself understood those visions to be visions. He

admitted even Byron and Hölderlin into respectable company. Nevertheless, he considered 'eccentricity,' as a 'departure from the common experience of the people' ([11], p. 149) or as the consistent deviation from accepted social norms, an indication of madness. For a moving document of the non-falsifiability of the judgment of insanity see Packard [40].

[34] Such a relation can be found implied in various medical attempts to define the character of a healthy mind. Conolly [11] speaks of a 'comparing faculty' which allows one to integrate what is sensed (or believed to be sensed) into the body of past experiences, or, one could paraphrase, to integrate what is immediately preceived and thus given to the mind, and what constitutes those assumptions about the world in our general experience, which allow us or forbid us to take that perception as representing a possible external reality (pp. 126, 337). Ray ([43], pp. 38ff, 79) considers the means–ends relation as that criterion, which can reveal whether the internal constitution of our knowledge is adequately mediated with the external world. Spitzka ([46], p. 60) assigns that function to a 'conscious ego' as a collection of past experiences plus a functional disposition to add and incorporate new experiences – the latter here functioning as the 'external' and the former as the 'internal.'

[35] Or the capability to discriminate illusion from reality ([28], p. 37).

[36] Fools are not correctible: see Haslam ([28], p. 24); Spitzka ([46], p. 29); Spurzheim as quoted in Combe ([10], p. 195).

[37] See Georget ([21], p. 89); Combe ([9], p. 197); Ray ([43], pp. 110, 171); Beck and Beck ([4], p. 588). Fisher ([50], p. 156) calls them 'natural' causes. The only evidence for Henriette Cornier having been insane – in spite of the obvious 'premeditation' of her crime – consisted in a recent and unaccountable change in her character. ([21], p. 25). (For 'insane premeditation' see also Ray, [43], pp. 27, 39f.). No doubt, a 19th century Saul having changed into Paul would have been speedily secured by the authorities.)

[38] 'Zur Erkenntlichkeit würde der Arzt seinen Beistand dem Philosophen auch nicht versagen, wenn dieser bisweilen die grosse, aber immer vergebliche Kur der Narrheit versuchete. Er würde z.E. in der Tobsucht eines gelehrten Schreiers in Betrachtung ziehen, ob nicht katharktische Mittel, in verstärkter Dose genommen, dagegen etwas verfagen sollten. Denn da nach den Beobachtungen des SWIFTS ein schlecht Gedicht bloss eine Reinigung des Gehirns ist, durch welches viele schädliche Feuchtigkeiten zur Erleichterung des kranken Poeten abgezogen werden, warum sollte eine elende grüblerische Schrift nicht auch dergleichen sein? In diesem Falle aber wäre es ratsam, der Natur einen andern Weg der Reiningung anzuweisen, damit das Übel gründlich und in aller Stille abgeführet werde, ohne das gemeine Wesen dadurch zu beunruhigen' ([31], p. 315).

[39] Combe's ([10], p. 179) warning should not be overlooked: that 'continued excitement of *Ideality* leads to that endless, vague, and unattainable search after perfection, and restless dissatisfaction with ordinary views and arrangements . . . ' which must prove damaging to mental health. See also Battie ([1], p. 57) about '. . . the chimaerical dreams of infirm and shattered philosophers who after having spent many days and nights . . . in unwearied endeavours to reconcile metaphysical contradictions, to square the circle, to discover the Longitude or grand secret . . . have cracked their brains.'

BIBLIOGRAPHY

1. Battie, W.: 1758, *A Treatise on Madness*, J. Whiston, and B. White, London.
2. [Beccaria Bonesana, C.]: 1764, *Dei delitti e delle pene*, Monaco.
3. Beck, T. R.: 1811, *An Inaugural Dissertation on Insanity*, J. Seymour, New York.
4. Beck, T. R. and Beck, J. B.: 1838, *Elements of Medical Jurisprudence*, 6th edition, Thomas Cowperthwait, Philadelphia, 2 vols. (first edition: 1823).

5. Blackstone, Sir W.: 1899, *Commentaries on the Laws of England*, by T. M. Cooley, 4th edition, J. D. W. Andrews, Callaghan and Co, Chicago (first edition: 1765–69).
6. Broussais, F. J. V.: 1828, *De l'Irritation et de la Folie*, Melle Delaunay, libraire, Paris.
7. Chaillé, S. E.: 1876, *Origin and Progress of Medical Jurisprudence, 1776–1876*, Collins, Philadelphia.
8. Chapman, H. C.: 1892, *A Manual of Medical Jurisprudence and Toxology*, W. B. Saunders, Philadelphia.
9. Combe, A.: 1831,*Observations on Mental Derangement*, Longman, Rees, Orme, Brown & Green, London.
10. Combe, A.: 1834, *The Principles of Physiology Applied to the Preservation of Health, and to the Improvement of Physical and Mental Education*, 2nd edition, Adam & Charles Black, Edinburgh.
11. Conolly, J.: 1830, *An Inquire Concerning the Indications of Insanity*, John Taylor, London.
12. Diethelm, O.: 1971, *Medical Dissertations of Psychiatric Interest Printed Before 1750*, S. Karger, Basel, New York.
13. Elwell, J. J.: 1860, *A Medico–Legal Treatise on Malpractice, Medical Evidence, and Insanity, Comprising the Elements of Medical Jurisprudence*, Baker, Voorhis, New York.
14. Esquirol, J. E. D.: 1838, *Des Maladies mentales*, J.-B. Baillière, Paris.
15. Foucault, M.: 1961, *Folie et Déraison, Histoire de la folie à l'age classique*, Librairie Plon, Paris.
16. Foucault, M.: 1963, *Naissance de la clinique; une archéologie du regard médical*, Presses universitaires de France, Paris.
17. Friedreich, J. B.: 1833, *Systematische Literatur der ärztlichen und gerichtlichen Psychologie*, Th. Enslin, Berlin.
18. Friedreich, J. B.: 1836, *Historisch-kritische Darstellung der Theorien über das Wesen und den Sitz der psychischen Krankheiten*, Otto Wigand, Leipzig.
19. Friedreich, J. B.: 1852, *System der gerichtlichen Psychologie*, 3, umgearb. und verb. Auflage, Manz, Regensburg.
20. Georget, E. J.: 1825, *Examen médical des procès criminels des nommés Leger Feldtman, Lecouffe, Jean-Pierre et Papavoine*, Migneret, Paris.
21. Georget, E. J.: 1826, *Discussion médicolegale sur la Folie*, Migneret, Paris.
22. Georget, E. J.: 1827, *Des Maladies mentales, considerées dans leur rapports avec la législation civile et criminelle*, Cosson, Paris.
23. Georget, E. J.: 1828, *Nouvelle Discussion médico-legale sur la Folie*, Migneret, Paris.
24. Hale, Sir M.: 1800, *The History of the Pleas of the Crown*, S. Emlyn (ed.), London (first edition: 1678).
25. Hamilton, A. M.: 1883, *A Manual of Medical Jurisprudence*, Bermingham & Co, New York.
26. Hamilton, A. M.: 1894, *A System of Legal Medicine*, E. B. Treat, New York.
27. Hammond, W. A.: 1883, *A Treatise on Insanity in its Medical Relations*, D. Appelton and Co, New York.
28. Haslam, J.: 1807, *Medical Jurisprudence as it Relates to Insanity*, C. Hunter, London.
29. Highmore, A.: 1807, *A Treatise on the Law of Idiocy and Lunacy*, J. Butterworth, London.
30. Hobbes, T,: 1968, *Leviathan*, Penguin Books, Baltimore (first edition: 1651).
31. Kant, I.: 1922, 'Versuch über die Krankheiten des Kopfes', in E. Cassirer *et al.*,

(eds.), *Immanuel Kants Werke,* Bruno Cassirer, Berlin, vol. II, p. 301–16. (first edition: 1764).

32. Kant, I.: 1923, 'Die Religion innerhalb der Grenzen der blossen Vernunft', in *ed. cit.,* vol. VI, p. 139–354 (first edition: 1793).
33. Kant, I.: 1922, 'Der Streit der Fakultäten in drei Abschnitten', in *ed. cit.,* vol. VII, p. 311–432 (first edition: 1798).
34. Locke, J.: 1812, 'An Essay concerning Humane Understanding', in The Works of John Locke, 11th edition, W. Otridge and Son, London, vol. I, (Book II, Ch. XI, Sect. 13) (first edition: 1690).
35. Maudsley, H.: 1880, *The Pathology of Mind,* Appleton & Co, New York.
36. Maudsley, H.: 1867, *Responsibility in Mental Disease*, Appleton & Co, New York.
37. O'Dea, J.J.: 1886, 'Medico–legal science; a sketch of its progress, especially in the United States', in *Sanitarian,* 4, pp. 449–457, 493–503.
38. Ordronaux, J.: 1869, *The Jurisprudence of Medicine,* T. & J. W. Johnson & Co, Philadelphia.
39. Ordronaux, J.: 1878, *Commentaries on the Lunacy Laws of New York,* J. D. Parsons, Jr., Albany.
40. Packard, E. P. W.: 1975, *Modern Persecution, or Insane Asylums Unveiled,* Case, Lockwood & Brainard, printers and binders, Hartford.
41. Platon: 'Timaios', in 1871, *The Dialogues of Plato,* translated by B. Jowett, Oxford University Press, London, vol. III (original text in 1860: *Platonis Dialogi,* C. F. Hermann (ed.), Teubner, Leipzig).
42. Prichard, J. C.: 1837, *A Treatise on Insanity and Other Disorders Affecting the Mind,* Haswell, Barrington, and Haswell, Philadelphia (first edition: 1835).
43. Ray, I.: 1838, *A Treatise on the Medical Jurisprudence of Insanity,* C. C. Little and J. Brown, Boston.
44. Roscoe, W.: 1819, *Observations on Penal Jurisprudence, and the Reformation of Criminals,* T. Cadell and W. Davies, London.
45. Rush, B.: 1812, *Medical Inquiries and Observations upon the Diseases of the Mind,* Kimber & Richardson, Philadelphia (first edition 1794–98).
46. Spitzka, E. C.: 1887, *Insanity: Its Classification, Diagnosis and Treatment*, E. B. Treat, New York, (first edition: 1883).
47. Spurzheim, J.: 1833, *Observations on the Deranged Manifestations of the Mind, or Insanity,* Marsh, Capen & Lyon, Boston (first edition: 1817).
48. Tuke, D. H.: 1882, *Chapters in the History of the Insane in the British Isles,* Kegan Paul & Co., London.
49. Wharton, F., and Stillé, M.: 1860, *Treatise on Medical Jurisprudence,* Kay & Brother, Philadelphia (first edition: 1855).
50. Witthaus, R. A., and Becker, T. C.: 1906–11, *Medical Jurisprudence, Forensic Medicine and Toxology,* W. Wood & Co, New York, 4 vols.
51. Wyman, R.: 1830, *A Discourse on Mental Philosophy as Connected with Mental Disease, Delivered Before the Massachusetts Medical Society,* Daily Advertiser, Boston.

SECTION II

MENTAL ILLNESS AND MENTAL COMPLAINTS: SOME CONCEPTUAL PRESUPPOSITIONS

ALAN DONAGAN

HOW MUCH NEUROSIS SHOULD WE BEAR?

In his monograph, *Herbert Marcuse,* Alasdair MacIntyre has remarked that, while 'with Freud's own writings it is continually necessary for the reader to turn back from the theorizing to the case histories, from the inflated conceptual schemes to the revealing clinical detail,' Marcuse 'all too characteristically' has been primarily interested in Freud's metapsychology rather than in his method of therapy ([14], pp. 43–44). Although I endorse this judgment of MacIntyre's, the paper I am about to present is almost entirely metapsychological; that is, it is about certain concepts of mental disease and illness, and not about what those concepts have been (rightly or wrongly) applied to. However, my purpose has been clinical. I have concerned myself with the implications of certain metapsychological doctrines for psychiatrists' attitudes to their patients; and, indeed, for anybody's attitudes to himself or to anybody else.

The fundamental topic I propose to investigate is commonplace: namely, whether the concepts of disease and illness must be modified or transformed when their application is extended from the human body to the human mind. We do not have to be Cartesians to think that, as human agents – as active participants in a civilized human life -- we are more intimately identified with our minds than with our bodies. Apart from congenital defects and weaknesses, defects in the functioning of our bodies are in large measure (although not entirely) things we must put up with, but defects in the functioning of our minds are for the most part (although again not entirely) a matter of what we do. Shortness of breath and stiffness in the joints are *sufferings;* instability of purpose and rigidity of opinion in face of adequate contrary evidence are bad dispositions of *action.* Here I take actions to be either bodily or purely mental processes that are intentional under some description.[1] And, although the point is a subtle one, bodily suffering is always a matter of bodily processes as processes, and not as actions. A person who suffers from a tottering gait has a malfunction in his legs or back in virtue of which, when he walks, he totters. He suffers inasmuch as the malfunction impairs his capacity to walk, not inasmuch as he walks. But mental disease is *prima facie* not like this. In themselves, the actions of neurotics and psychotics are like those of anybody else, but they are inappropriate to their situation. The question is

H. T. Engelhardt, Jr. and S. F. Spicker (eds.), Mental Health: Philosophical Perspectives, 41–53.

whether a concept of disease applying primarily to bodily malfunctioning can be extended to conditions in which agents act inappropriately, without transforming its character.

This topic suggests another which is more common in the literature of mental disease and illness. In itself, bodily disease is a human evil, even though it may accidentally be to a patient's benefit (he would not have read *War and Peace* unless he had been a convalescent), or may be the occasion of a triumph of character over misfortune. Nobody, I think, has ever held that bodily disease is a necessary condition of any characteristically human achievement. It is quite otherwise with mental disease. This, I take it, was in W. H. Auden's mind when, in his poem *New Year Letter,* he offered advice which lodged in my mind for thirty years to give this paper its rather ridiculous title:

> Give to each child that's in our care,
> As much neurosis as the child can bear.

Fifty years before, Nietzsche had given a related warning:

> Health and sickliness: one should be careful! The standard remains the efflorescence of the body, the agility, courage, and cheerfulness of the spirit – but also, of course, how much of the sickly it can take and overcome – how much it can make healthy. That of which more delicate men would perish belongs to the stimulants of *great* health ([15], sect. 1013, p. 523).

That Nietzsche had in mind primarily sickliness and health of the spirit is sufficiently obvious.

In most contemporary treatments, a generic concept of disease is elicited by abstraction from the concept of bodily disease, and mental disease is then taken to be a species falling under that genus. Even many, like Nietzsche, who have taken some degree of mental sickness to be a necessary condition of the highest cultural achievement, have followed this course. In order to investigate the problems raised by doing so, I shall begin by outlining the concept of mental disease to which it leads. And, since it will be convenient to illustrate that concept by an example of an actual theory constructed in terms of it, I have chosen Frued's for the purpose, because it is familiar. It will become evident that, for the kinds of question I shall raise, the merits and demerits of Freud's theory by comparison with either its successor or its rivals in its own time are of little importance.

Bodily illness is usually conceived as a discomfort or malfunction arising from a deviation from the normal form or functioning of the body which, although it may be latent, tends to cause illness.[2] So conceived, physical disease presupposes that the normal human body has a certain form (so that, for example, congenital conditions such as having a harelip are judged deformities and are medically considered to be diseases), and that it functions in a certain way.

It is now usual to think of the functions of the various systems that compose the human body as homeostatic: that is, as maintaining in a certain normal inner condition. Here a distinction must be drawn. Some of the body's homeostatic systems have to do with keeping its inner state normal in face of a constantly changing environment (e.g., in respect of temperature, light, moisture). These are homeostatic in the primary sense. Others have to do with protecting the primary homeostatic systems, whether from stress or shock, or from attacking microbiological agents, and with restoring their normal functioning when they have been damaged. Since these systems are indirectly concerned with maintaining the body's normal inner state, although not directly, they may be accounted homeostatic in a secondary sense. Any deviation from normal functioning in any of these homeostatic systems, whether primary or secondary, threatens the preservation of the normal inner state of the body, and is therefore, medically speaking, a disease.

A generic concept of disease, applying to non-bodily as well as to bodily diseases, is readily abstracted from such a concept of bodily disease. All that need be done is to lay it down that a human being is composed of primary and secondary homeostatic systems, some bodily and some not; and to define disease in general as a deviation from normal form and functioning that tends to cause discomfort or malfunctioning. And mental diseases may be brought under this generic concept by supposing that the human mind, like the human body, has a certain normal inner state, and is composed of various homeostatic systems, primary and secondary, having to do with the maintenance of that state.

This generalization of the concept of disease obviously extends the concept of what is sometimes called the 'disease entity.' If it is accepted, then disease entities that are mental must be recognized. But no commitment to philosophical dualism is entailed. It is not denied that mental events and processes may ultimately turn out to be identical with physical events and processes, as materialist identity theorists contend. Freud himself professed to be a materialist; but that did not deter him from developing a theory of mental disease of which none of the fundamental concepts are physical. The denunciations of 'Cartesian dualism' now customary in the philosophy of medicine seem to be a ritual the meaning of which is obscure. They do not imply that mental disease is to be described in physical terms. (That would be 'reductionism,' which is also ritually denounced.) A dualism of description, then, is not only tolerated, but insisted upon. That is the only dualism presupposed in the positions with which I am concerned.

The theory of mental disease in Freudian psychoanalysis is a good example of how such a concept may be given specific content. In his own *Short Account of Psychoanalysis,* written in 1923, Freud summed up his theory, of which he asserted that it 'appeared to give a satisfactory account of the origin, meaning and purpose of neurotic symptoms,' as follows:

> [E]mphasis on instinctual life (affectivity), on mental dynamics, on the fact that even the most obscure and arbitrary mental phenomena invariably have a meaning and causation, the theory of psychical conflict and of the pathogenic nature of repression, the view that symptoms are substitutive satisfactions, the recognition of the aetiological importance of sexual life, and in particular of the beginnings of infantile sexuality. From a philosophical standpoint, this theory was bound to adopt the view that the mental does not coincide with the conscious, that mental processes are in themselves unconscious and are only made conscious by the functioning of special organs (agencies or systems). By way of completing this list I will add that among the affective relations of childhood the complicated emotional relation of children to their parents . . . came into prominence. It became clear that this was the nucleus of every case of neurosis ([10], pp. 197–198).

The principal homeostatic function presupposed in Freud's account of mental dynamics is the direction of the flow of instinctual energy to acceptable discharge. Instinctual energy is taken to be directed towards a specific kind of discharge with respect to specific objects. When somebody is consciously averse to the behavior required for such a discharge, and so does not engage in it, the energy directed towards it is dammed up. The dam must ultimately give way, as more and more energy presses against it; but (and this was Freud's theory of repression) it is possible to ensure that the dammed up energy will not be discharged in a way deemed unacceptable if the very existence of a thrust towards such discharge is kept from consciousness.

Even repressed and therefore unconscious energy will in the end force its way to discharge; but that discharge will not be what it would have been had awareness of its original direction been allowed into consciousness. It may be redirected towards a form of discharge which the person in question finds acceptable and even desirable, as in the process known as sublimation. But if the repressed energy is not acceptably redirected, it will force a discharge in substitute ways which are not closed by repression, and which the agent will experience as undesired symptoms. In Freud's view, homeostasis is maintained provided that the flow of instinctual energy is regularly discharged either directly or in a sublimated form, so that undischarged energy never accumulates in such quantity as to force a discharge that is undesired and symptomatic.

However, even at the highly general level of this account, Freud's concept of mental disease diverges in an interesting way from that of its bodily analogue. While the concept of the human body's normal form and functioning is in no way socially determined, that of the normal functioning of the mind is, inasmuch as it is defined in terms of socially acceptable discharge. It is tempting to see this as a defect, and to try to correct it by working out a purely 'natural' conception of the normal functioning of the human mind, which may be used in criticizing the character of social institutions. This was done by the so-called 'left' Freudians, and above all by Wilhelm Reich.[3] And a few years ago, Erich Fromm called for a radical humanist psychoanalysis 'concerned with the processes that could lead to the adaptation of society to the needs of man, rather than man's adaptation to society' ([11], p. 29). Such a theory presupposes that the normal or natural functioning of the human mind can be determined independently of the actual demands upon it made by this society or that.

Freud distinguished three agencies or systems as together comprising the human mind: namely, id, ego, and superego. The fundamental one is the id, a system of energies in themselves unknown and unconscious, seeking discharge. '[T]he pleasure principle . . . reigns unrestrictedly in [it]' ([7], p. 25), according to Freud, by which I take him to mean that all the functions of the id are different forms of accumulating and discharging affective energy, the discharge being experienced as pleasant, and its prevention as unpleasant. Within the fundamental id, there arises in the higher animals, but most markedly in human beings, a perceptual-conscious system, which Freud held to comprise not only sensation (both inner and outer), but also the use of language (which he called 'word-presentation'), and to constitute the nucleus of a second system of psychical energy, the ego ([7], pp. 20, 23). The ego is a system involving awareness of the body and physical environment of the human being whose ego it is, and also of the various energies of its id (of course, a limited awareness). Accordingly, Freud wrote, in a metaphor that is natural and vivid, 'The ego is not sharply separated from the id; its lower portion merges into it' ([7], p. 24).

Two of the functions of the ego in Freud's theory may be singled out. The first is that of 'forcing itself, so to speak, on the id as a love-object,' and thus transforming 'object-libido into narcissistic libido' ([7], p. 30). The second is that of 'bring[ing] the influence of the external world to bear upon the id and its tendencies' ([7], p. 25): in other words, of modifying the id's pressure for discharge of its accumulated energies by an awareness of what is possible in the world in which it exists. Thus 'the ego represents what may be called reason and common sense, in contrast to the id, which contains the passions' ([7], p. 25).

In carrying out the second of its functions, the energy of the ego is directed to thwarting the direct discharge of the energies of the id. In part it does its work consciously, as when a man deliberately refrains from satisfying some impulse of which he is aware. But the ego neither can nor does rely wholly on the energies connected with consciousness: 'In its relation to the id,' Freud observed, 'it is like a man on horseback, who has to hold in check the superior strength of the horse; with this difference, that the rider tries to do so with his own strength while the ego uses borrowed forces' ([7], p. 25).

Freud's theory of how the ego borrows from the id forces to control the id is complicated, but at bottom it is this: Each individual's character is first decisively formed when the Oedipus complex is 'demolished' or 'dissolved' by forming a 'precipitate' in the ego, consisting of two identifications, one with the father, and one with the mother, which are combined. These identifications occur because each individual is bisexual, and because the Oedipus complex itself is a three-termed relation, in which attachment to the parent of the opposite sex, and the Oedipal rivalry it engenders, is accompanied by attachment to the parent of the same sex, the Oedipal rival. Rivalry with the parent of the same sex is controlled by taking into the ego, in the form of an 'ego-ideal' or 'super-ego,' the strength and the authority of that parent, with the result that all consciousness of the Oedipal striving of the id is suppressed ([7], pp. 31–35).

Freud summed up the whole process as follows:

> The ego develops from perceiving instincts to controlling them, from obeying instincts to inhibiting them. In this achievement a large share is taken by the ego ideal [i.e., the superego], which indeed is partly a reaction formation against the instinctual process of the id. Psycho-analysis is an instrument to enable the ego to achieve a progressive conquest of the id ([7], pp. 55–56).

In other words, in Freud's mentally healthy human being, the functions carried out unconsciously by the superego are brought to light, and as far as possible transferred to the ego. What the ego was too weak to do without borrowed forces in childhood, it is to be enabled to do alone in maturity.

The crucial point in Freud's view of mental disease and mental health, as in its lineal descendants and principal rivals, is that it treats the actions of human beings as a straightforward product of the interaction of the various physical and psychical systems of which they are taken to be composed. What Freud calls 'ego' in a human being, that is, perceptual and intellectual consciousness, and the energies it generates, is much stronger and more comprehensive than its counterpart in brute animals, but a human being is not his ego. He is a psychophysical complex, in which ego is one system among others; and what that complex does is no more the work of the ego than it is of any of the other interacting systems. Freud's metaphor of conquest was felicitous. In states founded on conquest, the conquerors depend on the work of the conquered for their subsistence.

Freud's use of the term 'ego' (in German, 'Ich'), and his numerous statements in which it figures as a quasi-agent, may have concealed from him, as it has from many of his readers, that along with the radical departures his theory makes from the traditional conception of a human being to which he did draw attention, there is another, equally radical, to which he did not. Traditionally, brute animals were regarded as complex systems, the activities of which are the product of certain drives and desires on one hand, and certain perceptions on the other, and which possess capacities to modify kinds of activity in which its drives are thwarted or its desires unsatisfied. Human beings were considered to be different in kind. They were held to be endowed with intellect; and intellect, through which alone mastery of language was considered possible, was not thought to be a mere extension of perceptual consciousness, as apparently it was by Freud. An animal endowed with intellect was held to be an individual of a different and higher kind of any brute animal, because it is an agent in a different and higher sense. For the activities of an intellectual creature are not simply products of its drives and desires, and of its perceptions. It is both able to conceive ends towards which it has no felt drive or desires, and to choose to act in view of those ends. It is, in short, capable of autonomous choice. It is endowed with *will*. Such a view of man, as differentiated from other animals, was almost completely worked out by Aristotle; and (to confine ourselves to philosophers of the very first rank) despite numerous differences on other matters of importance, was adhered to by Descartes and Kant.[4]

This traditional conception of the human mind is plainly incompatible, not only with Freud's theory of mental disease, but with any other which takes mental disease to be analogous to bodily in the way we have described. By definition, a homeostatic system is one whose activity is determined by the end to which its functioning is directed. No homeostatic system chooses its ends for itself: in other words, no homeostatic system is autonomous. It therefore follows that if what is fundamentally human in human beings is an autonomous intellect, human beings cannot be simply complexes of primary and secondary homeostatic systems. It further follows that, if strictly speaking there are mental diseases at all, they must be something other than deviations in homeostatic systems from their normal form and functioning.

In addition, but derivatively, Freudian theory diverges from the traditional conception of man in the connection it finds between what it calls 'ego' and a human being's identity as such. To Freud and his successors, the psychoanalytic ego takes the place of the mind or soul of traditional philosophy as what Erik Erikson has called 'the guardian of individuality.' Erikson has described the Freudian position accurately.

> Psychoanalysis [he writes] has not concerned itself with matters of soul and has assigned to consciousness a limited role in mental life by demonstrating that man's thoughts and acts are co-determined by unconscious motives which, upon analysis, prove him to be both worse and better than he thinks he is. But this also means that his motives as well as his feelings, thoughts and acts, often 'hang together' much better than he could (or should) be conscious of. The ego in psychoanalysis, then, is analogous to what it was in philosophy in earlier usage: a selective, integrating, coherent and persistent agency central to personality formation ([6], p. 147).

It is not unexpected that Erikson also finds something analogous to the psychoanalytic ego in brute animals, namely,

> ... a certain chaste restraint and selective discipline in the life of even the 'wildest' animals: a built-in regulator which prevents (or 'inhibits') carnivorous excess, inappropriate sexuality, useless rage, and damaging panic ([6], p. 150).

This is not the place in which to explore Erikson's doctrine that there is a mental illness which he has named 'identity crisis,' and which he has described as inability to establish one's station and vocation in life (Cf. [6], p. 64). He has provided abundant clinical evidence that in many individuals identity crises are connected with undeniably malignant symptoms and regressions. Yet he appears to be oblivious of the traditional conception of identity as simply human, and so overlooks the possibility that human identity, if there is such a thing, would be far deeper and more important than the potentially transient identity conferred by one's status and vocation – one's identity as a *fonctionnaire,* as Sartre might derisively say. It is perfectly possible to live normally as a human being without identifying oneself as having a certain status in a community, or a certain vocation; and anybody who does so live will be aware of himself as the same identical human being who may in his life pass from one status or vocation to another. According to the traditional view, to treat somebody whose regressive behavior was connected with an identity crisis, by helping him to identify himself as having a certain status and as following a certain vocation, would be symptomatic rather than radical. The radical treatment would be to help him to identify himself as a certain autonomous human being.

Despite the deep roots in popular consciousness of the traditional view of what a human being is, conceptions of the same general nature as the Freudian may be defended as required by modern philosophy as well as by modern science. In this paper I have dodged questions of the philosophy of science. Whether the traditional conception of man can be reconciled with modern science, and if so how, are complicated questions; and they are made more complex when the modern science about which they are raised is interpreted as modern science, as it will ultimately be. I have assumed that the practice of medicine must concern itself with human beings as they are disclosed to be, not by some philosophical interpretation according to the methods and principles of general theoretical sciences like physics and physiology, but by our historical knowledge of what human beings have done, among other things, in creating the sciences of physics and physiology, Quine's 'philosophy of science is philosophy enough' is not enough: 'philosophy of the history of science is philosophy enough' is nearer the truth.

When we turn to the conception of man in modern philosophy, we find that the traditional conception of him as an autonomous rational animal is assailed not only by classical empiricists like Hume, and their modern descendants, but also by radical phenomenologists like Nietzsche. The force of their attack, however, is weak. They correctly point out that the will is not an observable phenomenon. There are living human bodies, observable human behavior, including verbal utterances, and (although some deny even this) a non-material stream of consciousness, of sensation and feeling. The rest, we are told, is nothing but a survival of medieval religious notions. Of course, as Nietzsche pointed out in his *Nachlass*, a thoroughgoing application of criticism of this kind will dismiss as rubbish a good deal more than the human will: 'If we no longer believe in an effecting subject, the belief in the effecting thing collapses, as well as the reciprocal action of cause and effect between those phenomena that we call things.'[5] Against such a position, it is sufficient to point out that the fundamental concepts in terms of which both science and philosophy interpret the world are not empirical, although they are applied to observable process.

A concept like that of the will is not to be dismissed merely because it is non-empirical, any more than the concept of cause and effect. And the ground on which the concept of will is applied – on which human beings are considered to choose the ends by reference to which they act – is that only by doing so can we describe human beings as having done what we believe them to have done. I do not think it possible adequately to narrate the history of the psychoanalytic movement without ascribing to Freud something higher than a Freudian ego.

If, as I contend, a true theory of mental disease must incorporate the traditional conception of human beings as autonomous, what modifications will be entailed in such theories as the Freudian? The greater part of their structure could, I think, be left intact. In the traditional conception, it has never been questioned that human autonomy is exercised by beings who have numerous desires and impulses for the existence of which they are in no way responsible. Such a conception is deepened by adding to it the Freudian doctrine that these affective processes are in themselves unconscious, and are only made conscious by the functioning of other systems ([10], p. 197). Nor is there any objection to the Freudian conception either of the system by which affective energy is accumulated and discharged, as homeostatic or of the regulation of that discharge by repression in which the unconscious ego-ideal or supergo is the repressive force. The principal necessary modification is that it be acknowledged that a normal human being, when his intelligence has developed to a certain stage (what that stage is is a proper subject for inquiry), determines his actions according to ends which he himself chooses from among those he has become aware of as possible.

If this is so, Freud was mistaken when he likened the ego in its relation to the id to a man on horseback who has to hold in check the superior strength of the horse. Since a human being's desires, conscious and unconscious, are his own, they are not alien to him as a horse is to its rider. Nor will he, if normal, lack the power to refuse to perform any action at all by which a desire or impulse may be gratified. But it is a mistake to treat intellect as such as opposed to desire as such. It is a mark of rationality, when desire is strong and there is no reason not to gratify it, to act to gratify it. Because I invariably help myself to a drink when I am thirsty, a drink is available, and there is no reason not to take it, it does not follow that I lack the power to refrain. That I can refrain is shown by the fact that I sometimes do when I have reason.

Freud conceived mental disease as a deviation in the functioning of the three interacting systems whose function it is to maintain a regular discharge of accumulating affective energy. Accordingly, he defined a neurosis, etiologically, as a malfunction in which the ego, in its dependence on reality, suppresses a piece of the id, without redirecting it to acceptable discharge; and a psychosis as a malfunction in which, in the service of the id, the ego withdraws from a piece of reality ([8], p. 149; cf. [9], p. 183). In both cases, the homeostatic function of the id-ego-superego system is not carried out: accumulated affective energy is not acceptably discharged.

What modifications in these definitions are required by the conception of human beings as autonomous? Principally this: that the essence of mental disease be identified, not with any homeostatic malfunction, but with loss of control over a class of actions that are normally voluntary, or loss of normal power to arrive at the truth about oneself or one's situation, caused by the pressure of repressed affective energy.

As has been intimated, a neurotic does not lose the power to do or abstain from doing any particular action of which he is, in the ordinary sense, physically capable. Kleptomaniacs do not steal when they know a detective is watching them, obsessional neurotics perform their obsessive acts in such a way as to embarrass or inconvenience themselves as little as possible; and, short of collapse, phobics control their phobias when they believe they must. But nobody can maintain complete vigilance over everything he does in all its aspects. And any of the inevitable lapses of vigilance will allow repressed affective energy to find discharge in a neurotic act. A neurotic cannot overcome his neurotic impulses, obsessions, and fears as a class – he cannot completely control his tendency to neurotic behavior – even though he can control any individual impulse to act to which he pays attention.[6]

A similar modification can be made in any rival to Freudian theory in which mental disease is generally defined as a deviation in the form or functioning of the mind conceived as a complex of homeostatic systems, provided only that, like Freudian theory itself, the rival admits of modification to allow for the traditional conception of man as autonomous.

If neurosis is what Freudian theory, or any of its rivals, modified as I have proposed, says it is, it appears that the only reasonable answer to my semi-facetious question, 'How much neurosis should we bear?' is 'As little as we can contrive' – the very opposite of Auden's.

However, if it is assumed that Freudian theory, modified as I have proposed, is on the right lines, such an answer would be superficial. For, on that assumption, everybody is necessarily in the condition which, at a certain stage, gives rise to neurotic symptoms. Everybody is at least incipiently neurotic. And, according to Freudian theory, knowledge of this universal fact as it applies to oneself may take either of two forms: merely theoretical knowledge of a kind that may be gained by being supplied with a correct diagnosis of one's condition, and with sufficient general theoretical information to understand it; and what may be called 'effective' knowledge – the kind of knowledge of the repressive mechanism that has given rise to one's own neurotic symptoms that can be obtained only by experiencing a second time the conflict that was terminated by repression, which is possible only through a relation with somebody else to which the relevant elements of that early conflict can be transferred. If Freudian theory is correct, such transference is not confined to the analyst's consulting room. Erikson has argued that Freud's own self-analysis was made possible through a transference which occurred in his friendship with Fliess ([6], pp. 34–38). Nobody who has not neurotic symptoms severe enough to afford a compelling reason for attempting such a repetition of a painful conflict is likely ever to acquire that effective self-knowledge. And without it, one's knowledge of the general human condition will be – merely theoretical. Erikson stated this point well, if grandiosely, when he spoke of the first psychoanalyst's need to appoint 'his own neurosis that angel with which he must wrestle and whom he must not let go until his blessing, too, has been given' ([6], p. 23).

University of Chicago,
Chicago, Illinois.

NOTES

[1] The conception of an action as a process that is intentional under a description is due to G. E. M. Anscombe [1], and to Donald Davidson [4].

[2] As, for example, in *Encyclopaedia Britannica,* 15th edition (1974), s.v. 'Disease, Human' (by S. L. Robbins and J. H. Robbins), 'Psychoneuroses' (by H. P. Laughlin), and 'Psychoses' (by Silvano Arieti).

[3] Cf. Reich [17]. For a discussion of 'left-wing' Freudianism, see Fromm [11].

[4] For a discussion of the problems in Aristotle [2], see Hardie [12], Descartes [5], and Kant [13].

[5] Quoted and translated by Danto [3] from Nietzsche [15].

[6] This paragraph was written as a result of criticism by Arthur Diamond.

BIBLIOGRAPHY

1. Anscombe, G. E. M.: 1957, *Intention,* Blackwell, Oxford.
2. Aristotle, *Ethica Nicomachea,* III, chap. 2–5.
3. Danto, A.: 1965, *Nietzsche as Philosopher,* Macmillan, New York.
4. Davidson, D.: 1972, 'Agency', in R. Binkley, *et al.* (eds.), *Agent, Action and Reason,* University of Toronto Press, Toronto.
5. Descartes, *Les Passions de l'ame,* I, chap. 18, 31–34, 41–50.
6. Erikson, E.: 1964, *Insight and Responsibility,* Norton, New York.
7. Freud, S.: 1923, *The Ego and the Id,* in *The Standard Edition of the Complete Psychological Writings of Sigmund Freud,* trans. by J. Strachey, *et al.,* (hereafter cited as *Standard Edition*), vol. 19, 1961, Hogarth Press and Institute of Psycho-Analysis, London, pp. 12–59.
8. Freud, S.: 1924, *Neurosis and Psychosis,* in *Standard Edition,* vol. 19, pp. 149–153.
9. Freud, S.: 1924, *The Loss of Reality in Neurosis and Psychosis,* in *Standard Edition,* vol. 19, pp. 183–187.

10. Freud, S.: 1924, *A Short Account of Psycho-Analysis*, in *Standard Edition*, vol. 19, pp. 191–209.
11. Fromm, E.: 1970, *The Crisis of Psychoanalysis*, Holt, Rinehart and Winston, New York.
12. Hardie, W. F. R.: 1968, *Aristotle's Ethical Theory*, Clarendon Press, Oxford, ch. 9.
13. Kant, *Grundlegung zur Metaphysik der Sitten*, Abschneidung III.
14. MacIntyre, A.: 1970, *Herbert Marcuse*, Viking Press, New York.
15. Nietzsche, F.: 1967, *The Will to Power*, W. Kaufmann (ed.), trans. by W. Kaufmann and R. J. Hollingdale, Random House, New York.
16. Nietzsche, F.: *Aus dem Nachlass der Achtzigerjahre*, in K. Schlechta (ed.): 1958, *Nietzsches Werk in Drei Bände*, Hanser, Munich. pp. 540–541.
17. Reich, W.: 1933, *Charakter Analyse: Technik und Grundlagen*, Vienna: English trans., 1950, *Character Analysis*, Vision Press, London.

STEPHEN TOULMIN

PSYCHIC HEALTH, MENTAL CLARITY, SELF-KNOWLEDGE AND OTHER VIRTUES

I

In recent years, discussions about the concept of *mental health* have been much preoccupied with the influences – for good or ill – of something called 'the medical model'. The ways in which we talk or think about, and deal with, people who are afflicted of mind or spirit, and who seek out professional help in coping with these afflictions, have been increasingly guided (it is said) by patterns of thought borrowed from somatic medicine. In consequence we have been tempted to carry over, both in our attitudes towards the mental sufferer, and into the procedures that we employ in treating him, presuppositions and postures that are appropriate to his condition *only to the extent that there is a true analogy* between physical or physiological affliction/disease/therapy/cure/health etc., on the one hand, and mental or psychological affliction/disease/therapy/cure/health and the rest, on the other.

My aim in this paper is not to re-hash yet again the arguments that have been put forward in criticism, or in defense, of this medical or somatic model of mental health and/or disease. Rather, I believe the time has come to go behind the positions adopted in the current debate, and to try to get a better-proportioned view of the relations between mental and physical goods and ills; and also of the implications that these relations have for the actual conduct of our 'helping professions' – not merely the medical, psychiatric and psychotherapeutic professions, but also those other professions that involve the handling of complaints and the obtaining of remedies. For (I shall argue) in whatever terms we may seek to *distinguish* 'mental' from 'physical' and other disorders or afflictions, it will never be possible wholly to *separate* the human tasks involved in the professional handling of those different complaints; and any tendency we may have to insist on the distinctions in question may prove just as damaging in practice as the tendency to ignore them, which gave rise to the current debate in the first place.

I am ready to concede at the outset (that is to say) that there was a valid rhetorical point to be made a dozen or so years ago, when the contemporary attack on the medical model was warming up: e.g., in Thomas Szasz's denunciations of *The Myth of Mental Illness*. As a general topic, indeed, the

H. T. Engelhardt, Jr. and S. F. Spicker (eds.), Mental Health: Philosophical Perspectives, 55–70.

misusability of psychiatry is a recurrent one. (Chester Burns and Tris Engelhardt are full of splendid historical illustrations of this fact.) In this respect, the Soviet use of psychiatric hospitals as *loci* for the treatment of political dissidents – with the implication that this discontent with the current state of things in the Russian Empire is not a legitimate expression of rational preferences, but a mark of near-lunacy warranting hospitalization – is only a particularly flagrant and active instance of the same kinds of injustice and/or confusion that led people earlier to classify, say, masturbation, or a slave's impulse to run away, as 'psychopathological.' (Besides, the mote is in the other fellow's eye, so we can draw attention to it with becoming complacency.)

The point at issue has, none the less, been a *rhetorical* one, and more needs to be done to get it sorted out and articulated in ways that can stand up to analysis and criticism. Let me, at this stage, simply hint at the problems that need clarifying. A quick glance at the *Oxford English Dictionary* is enough to confirm that the category of 'ills' and 'illnesses' is *not* one that has been carried over in modern times, by some confusion of thought, from a primary, somatic sense to a metaphorical and misleading mental sense. On the contrary: from the beginning, 'ills' have been spoken of in quite general terms, as the opposite of 'goods,' and the noun 'illness' has been confined to the special sense of 'bodily disease' only in very recent years. There are, no doubt, important and interesting differences between 'ills of body' and 'ills of mind'; but it is not for nothing that they originally shared a common name, and we shall only distract ourselves (I shall conclude) if we treat the issue of mental illness as one about the nature and scope of *illness*, as such.

II

To focus at once on what I take to be the central burden of the discussion: those conditions that constitute 'illnesses' – properly so-called – are, on the current view, to be seen as disorders of the body, and are to be dealt with by direct treatment of the affected part or process. This being so, we need to distinguish clearly between those mental afflictions that are themselves side-effects of genuine bodily disorders, and those that have no such bodily source but spring rather from some kind of a moral conflict, behavioral maladaptation, or emotionally damaging habit of life. If we fail to do so, we may well end up by prescribing (e.g.) quite irrelevant and inappropriate drugs for complaints that involve no actual somatic defect; by masking the real source of the patient's unhappiness instead of revealing it; and in this way producing the semblance of a 'cure,' not by providing the patient with a genuine and relevant 'remedy' for his complaint, but by manipulating him, so that he no longer *feels* like complaining about his actual mental ills.

How are we to avoid this kind of manipulation? We should give up talking about 'mental illness' at all (it is said) except in those cases where there is real and specific disorder in, or damage to, the brain or central nervous system; we should acknowledge the essentially moral and behavioral character of all other 'mental' troubles; and we should be on our guard, as a result, against the danger of imposing our own moral and social attitudes on our clients – I say 'clients,' because even the term 'patients' may be prejudicial – under the pretense of giving them medical treatment. Otherwise, we are liable to slide insensibly from medicine into 'thought control.'

Now: I am of course quite familiar with the kinds of abuses against which this line of argument is directed. Anyone having the slightest acquaintance with the ways in which our prisons are run (for example) must know how tempting it is to cross the line between medical treatment and psychotherapy, on the one hand, and custodial constraint or control on the other. In any kind of institution, indeed, there may come a moment when the emotional burdens and afflictions of the inmates are more than the custodial staff can bear; and when a convenient diagnosis of schizophrenia can be used to justify a prescription of thiazine, and the tranquility it produces. (The other inmates may talk about this among themselves, not as 'healing the sick,' but as 'doping down the troublemakers'; but no matter, so long as the psychiatrist's conscience is clear.)

There is however a special problem to be faced in this kind of case, quite apart from the difficulties that arise over the concepts of 'mental health' and 'mental illness' in general. Any physician or psychiatrist who practices within an institutional framework has a *role conflict* to deal with. Whether he works in a prison, an asylum, or a home for the retarded, he is liable to be perceived – and even to perceive himself – as a member of the institution's administrative staff, not simply as a 'healer.' That being so, he cannot stand in the same relationship to those he treats as he would if he were in private practice. As inmates, they are in no sense his *clients*, and only in diluted sense '*his* patients.' First and foremost, they are the institution's *wards*, and the staff of the institution can clearly exercise a substantial degree of administrative latitude in carrying out their custodial responsibilities. (This is not to say that, in the eyes of the law, a prisoner or institutionalized mental patient loses all rights of action against a physician or psychiatrist for, say, battery; but the circumstances will normally be so coercive that there is little chance of such a suit being brought, far less sustained.)

Still: I do not want to get sidetracked on to the special problems that arise, and the special scope for the corruption of medicine or psychotherapy that exists, within these institutional settings. For the case against the idea of 'mental illness' has been stated in general terms, and it must be dealt with in general terms. So let me set those special cases aside, and return to the initial question: Is over-reliance on 'the medical model' essentially liable to lead to abuses and corruptions in other cases, equally – including those where the therapist has no direct concern with the client or patient, aside from that created by their contractual relationship?

To present my conclusions in advance, I shall argue, in return,

(1) that what has been called 'the medical model,' in the discussion of mental troubles and treatments, is only a special case of a more general 'mechanistic model,' which can be quite as damaging to the practice of internal somatic medicine as it is to psychotherapy;

(2) that the temptation of all medical complaints as springing from 'mechanical faults' in the patient's body creates the same opportunities for abuse in somatic medicine as exist in the field of mental health and afflictions;

(3) that it is scarcely ever possible to be sure, in advance, whether a patient's troubles are predominantly somatic, moral, social, or even legal, or a complex mixture of all kinds; and

(4) that we need to recognize that the *physiological* concept of a 'disease' or 'bodily defect' has always to be applied, in actual practice, within a broader *human* context, in which the primary operative concepts are not those of 'disease' and 'cure,' but rather those of 'complaint' and 'remedy.' On that human level, mental health and physical health can be considered on a par.

III

A full-scale social and anthropological history needs to be written about people's changing conceptions of disease. The general background of thought in any given culture has a significant effect on the ways in which 'falling ill' is perceived and dealt with. Sickness may be seen in one culture or epoch in terms of external agency and will, e.g., as a species of bewitchment: some other agent – human or divine – has 'willed' this affliction on me, and I must find some way of appeasing him, or of counteracting his influence on my condition. In another culture or epoch, sickness may be thought of in military terms: outside hostile forces have invaded my body and subverted its proper operations, with or without the help of betrayal by some Fifth Column within my body or spirit, and the task is to repel or neutralize the invaders.

(In its simpler-minded forms, the germ theory of disease continues this way of conceiving sickness.) In yet another, it may be thought of politically, as representing a breakdown of the institutional balance or power-structure within the Organism, in its analogy to a State, and the problem is to restore this political equilibrium. (Notice how the Organic Theory of the State can be played back, in this way, as a Societal Theory of the Organism.)

In our own days, all of these views continue to be influential in some quarters, and to some extent; but the dominant conception is yet a fourth one. This conception treats the workings of the body in mechanistic terms. Sickness is then to the internal organs and systems of the body what wounding is to its external parts: the breaking, wearing-out, or other malfunctioning of defective parts or systems within the Body-Machine. Correspondingly, therapy requires the repair, replacement, or restoration of function of the affected systems or parts. Whether or not this conception is actually owing directly to René Descartes himself, it has clear Cartesian affiliations, and its implications for our views about mental illness are plainly dualistic. Either our mental afflictions are by-products of bodily malfunctioning, in which case they are best treated by orthodox physiological repair-work; or else they are 'affections of the rational mind' – inaccessible to physiology or biochemistry, and of a different order from any physicochemical functions or malfunctions.

Each of these different conceptions of bodily functioning and sickness carries with it, of course, a correspondingly different conception of the 'doctor.' On one view, he will be an intercessor or spell-breaker; on another, a military ally or a diplomatic negotiator; on the third, a wise counselor or constitutional advisor; and, on the last, a glorified automechanic. In seeking medical advice, we shall in each case be seeking help of a different kind, depending on which way we perceive our condition. We may be in quest of reliable advice, either about how to deal with the external powers that have wished us this ill; or about how to repel the invasion and kill off the invaders; or about the establishment of a more secure and balanced bodily or psychic regime; or about getting our internal defects put right. Finally – and this is the crucial point for our present argument – each view will provide us with a different standard of comparison when thinking about mental afflictions. *Each view, in short, will serve to define a different 'medical model.'*

Now: how are we to decide between these four views? Some may find the answer to that question obvious. Surely (they will say) we know that the physiological/biochemical account of disease is the foundation of all effective modern medical therapy; so how could there be any doubt that the fourth view, and only the fourth, has any validity for us today? Given the extraordinary progress that has resulted over the last forty years in the development of antibiotics, the improvement of surgical procedures, and other techniques of medical intervention, it may seem churlish to challenge that conclusion. Yet, from here on, I shall be arguing that this position is – and can be – right only so far as it goes: right (that is to say) only in what it asserts, and not in what it denies.

Certainly, there are a great many sicknesses which can be thought of fruitfully as involving 'bodily defects,' as straightforward and plainly intelligible as the corresponding defects in a television set or an automobile. But that class of 'diseases' does not exhaust the types of complaint that the practicing doctor encounters in his consulting-room; and the 'bodily defect' story scarcely tells us the whole truth, even in the cases to which it most neatly applies. If it gives the appearance of being complete and exhaustive, in particular, it does so only at the price of deftly changing the subject at an early stage in the argument. We are persuaded by intellectual sleight-of-hand, at the outset, to trade in the everyday terms, 'complaint' and 'remedy,' for the more technical and highsounding terms, 'disease' and 'cure'; and the effect of this exchange is to conceal precisely those issues over which we should feel continued hesitations and reservations.

IV

By implication (I said) the 'bodily defect' or 'mechanical' view of disease is a Cartesian view. It sets – and cannot help setting – bodily disorders, including disorders of the brain and nervous system, over and against all those mental troubles whose diagnostic interpretation is not capable of being given clearly and directly in somatic terms. On the one hand, this feature of the view lends real plausibility to the argument that we should think about purely mental afflictions in quite different terms from authentic bodily disorders. On the other hand, any position that *takes for granted* a Cartesian-style dualism, and treats bodily and mental functions as *different in principle*, is so out-of-tune with the rest of our contemporary ideas that we should take a second look at its implications for our actual thought and practice before accepting it.

If we do so, some real and profound difficulties arise. In the first place, the Cartesian dichotomy of body and mind was designed – among other things – to justify keeping questions of *value* (whether logical or ethical) out of the world of *fact*. Valuation was essentially a task for the rational part of the human being, viz., the mind: meanwhile, all his bodily mechanisms went their own sweet way, through causal interactions that took place (or failed to take place) without benefit of 'reasons,' as the changes and chances of the physiological machine in fact turned out. Yet, in this respect, twentieth-century ideas about physiological 'functions' and their underlying bodily 'systems' have progressively abandoned any hard-and-fast distinction between 'facts' and 'values.' The healthy functioning of the respiratory system (for instance) is not just something we value positively because we happen to 'feel good' about it. Such healthy functioning is a self-evident 'good,' if not an actual precondition of any other real 'good.'

I shall not elaborate on that point here, since I dealt with it at length in the paper I delivered at the previous Galveston meeting in this series.[1] In any event, there are further, more basic difficulties to be dealt with, when we turn to consider the relevance of the 'mechanical' model to the actual practice of medicine. But there is just one thing that can usefully be said here about the notion of 'mental functioning.' The more complex and subtle physiological functions of the human frame, beginning with the vasomotor and other homeostatic functions analyzed by Claude Bernard, are not mediated by *single* parts or organs of the body: instead, they involve complicated and highly-interlinked *networks* of parts, that react back on one another according to negative feedback and other control principles.

When one of the most complex functions of all (e.g., the visual system) is disturbed, it may well be impossible to pin that functional disorder down to any specific 'defective bodily part.' Rather, we may have to conclude that the entire system in question is – for some possibly transient historical reason or other – temporarily maladapted to those elements in the environment with which it normally interacts. To put it bluntly, I may be unable to focus my vision, because I am drunk, or have been overworking. And, more generally, the more we move across the spectrum that runs from simple, local 'bodily events,' by way of the homeostatic systems, to the progressively higher 'mental functions,' the more we have to view the bodily frame not as a mechanical collection of parts, but as a complete, integrated system in adaptive interaction with its environment. In addition to mechanical defects of the *parts*, there are accordingly functional defects of the *whole*, and also – more particularly – higher functional defects affecting the interaction between the entire organism and its external environment.

Far from having to set the world of mental activities and adaptations quite aside from that of bodily processes and functions, therefore, we are at liberty to see them as so many points along a continuous line. There are indeed some functional disorders that arise from local bodily defects: e.g., a failing valve in a patient's heart. But there are also others that spring from internal maladaptations, or misalignments, between the different sub-systems in a given physiological system, or between entire systems: these faulty interactions may result, e.g., in tachycardia. And there are yet others that arise from failures of adaptation – not least, failures to *learn* an effective way of adapting – to crucial features of the external situation: a term which includes (of course) both other people and also the effects of the patient's own conduct. While the higher mental functions of the human frame may thus be distinguishable from its local bodily processes, it does not follow that they are separable from them, or that they can usefully be dealt with in complete isolation from them.

A similar conclusion follows if we consider the complex interlinked syndromes in which mental and bodily troubles commonly present themselves in actual practice. When a patient enters the doctor's consulting-room (as has often been said) he does not yet have a *disease*: he has a *complaint*. And one of the first questions that arises for the doctor is, how far this complaint springs from an actual bodily defect, how far from confused emotions or a disordered life. The patient suffers from shortness of breath; and this is evidently a side effect of obesity. Yet is that obesity, and the unhappiness associated with it, a sign in turn of glandular dysfunction? Or is the overeating by which the obesity is maintained a response (maladaptive, maybe, but an intelligible response nevertheless) to that unhappiness? So long as we stay on this first pre-diagnostic level, of 'complaints' rather than 'diseases,' we can remain open in our attitudes towards the psychophysiological complexities of actual human affliction.

In theoretical terms, then, a 'disease' may be a perfectly well-defined entity, of which more or less clear instances are to be met with in actual practice. In practice, however, it is never a direct object of observation, but at best what Max Weber called an *ideal type*. The presence of a specific disease, or bodily disorder, may play a large part in explaining a patient's complaint; but it is the complaint that he comes in with, and the complaint that has to be dealt with. So, in medical practice, all complaints have multiple aspects – somatic, physiological, social, even environmental – and good medical treatment is concerned with all these aspects at once. If we cannot help our obese patient (for example) to deal with his unhappiness well enough so that he in fact takes the medication he needs to bring his glandular troubles under control, how shall we be able to deal successfully with either facet of his troubles? For us to think exclusively in terms of the mechanistic, or bodily defect model of sickness can therefore be quite as damaging to our understanding of normal somatic medicine as it ever is in psychotherapy. It may even be more damaging in the somatic case, perhaps, since in psychotherapy the 'mental' component is unmistakable, whereas in internal medicine the mechanistically-minded physician may be tempted to dismiss all the patient's psychological symptoms as signs of stupidity and immaturity, rather than as integral elements in his complaint.

We are now in a position to see in better proportion what lies behind recent objections against the influence of 'the medical model' in psychotherapy. Firstly: these objections have been directed not against *the* medical model, since there is no single, well-defined 'medical model.' Rather, they are directed against the mechanistic, or bodily-defect model, which is only one out of at least four alternative 'medical models.' Secondly, this particular mechanistic model is no less harmful in regular medicine than it is in psychotherapy. It may indeed be dehumanizing and corrupting to treat a mentally-afflicted patient as a defective machine, and to resort to drugs or other bodily interventions instead of having proper respect for his moral, rational, and emotional difficulties. But it can be just as dehumanizing and corrupting to treat a physically afflicted patient in the same way. Whatever the predominant causes of the patient's complaint may be – whether they lie principally within his body, or principally in his interactions with the world of his daily life – the task of treating him is best dealt with on all levels at once; and, by adopting an over-mechanistic approach, the physician may frustrate his own best efforts to deal with the problems in hand. He may fail to cure the *disease*, that is, simply through mismanaging the *patient*.

Conversely: just as the good physician can avoid being over-mechanistic in his approach, and seek to establish a good human relationship (or 'therapeutic alliance') with the patient, so too the medically-oriented psychotherapist is free to do the same. It is not his acceptance or rejection of medical analogies that makes the crucial difference to the psychotherapist's attitudes: rather, it is his readiness to accept the patient on an equal human footing, instead of approaching him in a manipulative or authoritarian spirit. And in this respect (it seems to me) the situations in internal medicine and psychotherapy are, once again, much more alike than they are different.

Parenthetically: there is something of a familiar American kind about the temptation to take up this mechanistic attitude towards disease. It has often been remarked that Americans like to think of all difficulties and obstacles as 'problems' calling for 'solutions,' and that they are temperamentally unready to accept them as standing conditions which may just have to be lived with and dealt with day by day, year by year. This tendency shows itself in politics, in diplomacy, and in a dozen other fields besides medicine. (Surely, for instance, Kissinger should be able to find some way of 'solving' the Middle East 'problem?') If this is so, it certainly helps us to understand why Americans should find the concept of 'disease' so much more congenial than that of 'complaints.' For, just as problems have 'solutions,' diseases have 'cures.' Complaints, by contrast, may simply have to be managed, day by day, year by year, until death closes the book; and the best that we can hope for in that case is some more-or-less adequate, more-or-less effective, more-or-less lasting 'remedies.'

In addressing an audience of physicians, I feel a certain diffidence about talking in abstract, theoretical terms about matters that touch them personally every day of their working lives. For, of course, no practicing physician with any capacity to think and feel can possibly be an active party to the Great American Conspiracy to deny the fact of Death. If the central job of medicine were really to cure diseases, then a competent medical profession would have abolished death long ago. Yet terminal illnesses are illnesses just as much as any others: they are merely ones in which the range of symptoms for which there are available remedies is more restricted than usual, and those remedies themselves are less lasting in their effects. To the practicing clinician, then, the *management* of the sick patient is the primary fact of life, as well as (more importantly) the primary fact of death. And this task of management is one in which 'mental' and 'physical' aspects – the correction of bodily defects and the alteration of the patient's mode of life and self-perceptions – have an equal part to play.

V

In thinking about the problems of mental suffering and mental health as much as about somatic medicine, therefore, we should be prepared to set aside the ideas of 'disease' and 'cure,' and look more closely instead at the twin notions of 'complaints' and 'remedies.' For whatever the nature of the patient's condition – whether its main focus is some physiological disorder or some psychological confusion – good clinical care demands a kind of management which pays proper attention to both aspects, and does so in a way that avoids the simplistic, problem-solving, auto-mechanic approach.

Let us bring this discussion back to the issue of mental afflictions, in particular. Diseases (I said) have 'cures,' but complaints have 'remedies.' Cures are either efficacious or inefficacious; but remedies have all kinds of 'virtues.' From the point of view of the present paper, then, the basic questions to be asked about mental afflictions are: firstly, 'What kinds of *remedies* do they lend themselves to?,' and, secondly, 'What kinds of *virtues* will those remedies characteristically have?'

When stated in these terms, the *philosophical* problems raised by the concept of 'mental health' are neither particularly difficult, nor particularly unique. If the physician concludes, for instance, that a fat patient's obesity is more owing to his unhappiness (and that in turn to a bad habit of undervaluing himself) than the unhappiness is owing to the obesity (and that in turn to some bodily disorder), he may rightly decide that the best remedy for the patient's general condition will come from psychotherapy rather than, say, hormone treatment. But this does not imply that the physician is thereby ceasing to practice medicine, or to pay attention to the patient's *medical* complaint: on the contrary, he is merely seeking a remedy for that complaint in a different direction.

Of course, the *kinds* of virtues that these 'mental' remedies will have for the patient are as different from the virtues of (e.g.) an efficacious antibiotic as the original 'mental' difficulties are from bodily disorders or defects. The outcome of good psychotherapy will show up, not primarily in bodily changes like the disappearance of fever or the healing of lesions, but in a new mode of life, and in more realistic self-perceptions. Hence the title of my paper: mental complaints, their management and treatment, the remedies they demand, and the virtues those remedies can have, share one thing in common. They have as much to do with the patient's own confused emotions, his self-appraisals, and his ways of dealing with his family, his associates, and his human relations generally, as they have to do with his glands, his circulation, or even his central nervous system. As such, mental difficulties may often be somewhat intractable, and resist any swift 'cure.' Rather, they may tend to be standing conditions, which call for intelligent and sensitive management, quite as much as for a quick fix. (However much psychotherapy may help, I would question whether we can ever hope to be entirely 'cured' of our psychological vulnerabilities: rather, we may learn to understand them, and to control their consequences better than we did before.)

Yet are we any the worse off for that? The chance of achieving a new kind of self-understanding or self-knowledge, the chance of clearing up the confusions in our own self-perceptions and self-appraisals, the chance of coming to 'know our own minds' and so to act more effectively or from less conflicting intentions: all these things are surely worth having and – in some cases, at any rate – they may well be just the remedies that our initial complaints really call for. I say, 'in some cases,' for good reason. Earlier on, I argued that the mechanistic conception of disease, as a 'bodily defect,' refers not to an empirical entity but to an explanatory ideal type; and the same is true, equally, of the concept of a mental illness. Often enough, indeed, our afflictions spring from a mixture of failings, mental and physical. Obesity due originally to a glandular imbalance in childhood may provoke self-disgust, which leads to compensatory overeating, so aggravating the initial obesity, which reinforces the self-disgust . . . and so on. Such a complaint can be treated satisfactorily, only by seeking remedies of several kinds at the same time. Why, then, should we say that such a condition is *either* basically somatic *or* basically mental, when it is clearly a bit of both? A proper treatment, in such a case, will correspondingly have to mix physical and mental remedies: a bit of hormone therapy, maybe, but also a better regime and diet, and a chance to talk out the emotional aspects of the difficulties – to engage in what Anna O. herself described to Joseph Breuer as personal 'chimney-sweeping' and 'the talking cure.'

VI

Considered merely as an abstract concept, then, the category of 'mental disease' can indeed be made to look pretty paradoxical, at any rate, to the residual Cartesian lurking in us all – especially if we think of diseases as 'bodily disorders,' and so turn mental diseases into 'mental bodily disorders.' But that is only an abstract point. If considered rather within the context of actual practice (I have been arguing) the category of 'mental illness' is anything but a myth. For the primary reference of the term 'illness' is to human affliction, suffering, and pain – in short, to the whole range of human 'ills' – and so to *complaints* rather than to *diseases*. And the range of complaints with which patients enter a doctor's consulting room covers an entire spectrum of bodily, mental, and mixed complaints, ranging from the purely somatic (simple infections and broken bones) at one extreme, to the purely mental (anxiety neuroses and personality disorders) at the other.

What I am seeking to underline here is the continuity of that spectrum. In the actual life of the practicing doctor, the pure 'physician' role and the pure 'psychotherapist' role are quite as much *ideal types* as the notions of 'disease' and 'mental illness' themselves. If a doctor is to get his patient the remedies he needs for his actual condition, he cannot afford to be picky about which of these roles he performs: he must be prepared to be a bit of a physiological mechanic, a bit of a personal counselor, a bit of a confessor, dietitian, and even home economist. How the patient is living, how he perceives his own situation, feels about his own capacities, and so on, may have just as much bearing on how he copes with (say) glandular insufficiency, or a chronic heart condition, as all the drugs in the pharmacopoeia. To put the point tritely: what the patient primarily needs is not a 'cure' – for we cannot be sure at the outset that he has any specific 'disease' – but rather medical *attention*. And if his current complaints are *attended to* in the ways that their true character calls for, that may require from the doctor not just a dose of physick, but equally a little bit of talking, and a great deal of listening.

Having reached this point in my paper, I want to broaden the canvas a little, and make some final remarks about the nature and problems of *professionalism*. In talking about the continuities that link mental and physical ailments, I have deliberately chosen to use the terminology of complaints and remedies, rather than that of diseases and cures; and it may have struck some of you that the terminology of complaints and remedies is also the language of *legal* practice. This is no accident. For the remarks I made earlier, about the impossibility of telling for certain in advance whether a patient's troubles are predominantly physiological or psychological, can be generalized; and the arguments for seeing a continuity between somatic medicine (together with the 'physician' role) on the one hand, and psychological medicine (together with the 'psychotherapist' role) on the other, can usefully be extended further.

The fact is that, when a patient enters a doctor's consulting room in a state of distress, there is no way that the doctor can tell for certain in advance *even* whether this is truly a 'medical' case, in the broadest sense – including psychological and somatic medicine under a single heading – at all. A woman with two young children and little professional training loses her husband in an automobile collision, and goes to her doctor displaying evident signs of stress: headaches, insomnia, digestive difficulties, and so on. She is evidently right to seek professional help; but who can really be sure that her problems are basically medical problems, rather than (say) legal problems or welfare problems? Maybe she needs help to recover damages from the driver of the other car; maybe she needs to find home help with the children, and job training so that she can become self-supporting; maybe she would profit from the spiritual counsel of a wise priest or minister; and, then again, maybe a short sharp course of psychological counseling and/or tranquilizers will do no harm in the meanwhile. What, then, does the doctor do? Does he simply give her a physical check-up, tell her that there is nothing physiologically wrong with her, and show her out? Does he take it for granted (that is to say) that, when she consulted a doctor rather than a lawyer, a priest, or the social security office, *she knew what she was doing*? Does he limit himself, in short, to the 'physician' role, with or without a dash of the 'psychotherapist' role, for which he has had actual professional training?

I shall not be answering those questions directly here. Clearly, different doctors respond to such personal crises in different ways. All I do want to say is this: that our present era of professional specialization is liable to bear down very hard on the people who really need help. Someone who is in shock, as a result of finding herself placed in the kind of position that I have supposed, is simply in no state to know *whom* to turn to for help. She may end up by going to the doctor simply because he happens to be there on the spot, while the nearest lawyer or welfare officer is in the next town ten miles away. This being so, it will surely be quite natural and appropriate for the doctor to mix in welfare, legal, and/or spiritual counsel, together with his medical and psychotherapeutic advice. For, given the patient's actual situation, she may have a legitimate 'complaint,' not merely in a medical and/or psychological sense, but also in other respects – e.g., in legal respects. The 'remedies' she should be advised to look for may thus include legal, as well as medical and psychotherapeutic remedies: e.g., damages by way of legal redress from the other motorist, whose reckless driving caused her husband's death. And, though the doctor may quite rightly see that he is unqualified to help her with the entire process of obtaining those legal remedies, he will surely do his own job better if he explains to her, not only why and how she should adjust her diet and medication to deal with the medical symptoms of shock, but also why and how she should go about finding a lawyer who can help her to formulate her legal complaint, and obtain the remedies she is entitled to on that other account.

VII

Let me sum up. I began this paper by discussing the argument that the problems of mental health and mental illness are radically confused by the influence of 'the medical model.' I concluded that this argument is directed at the wrong target. It is not 'the medical model' in general, but one particular mechanistic conception of 'disease' that is at fault here; and the damage it can cause is not confined to psychiatry and psychotherapy. In actual practice (I argued in return) there is a genuine continuity between mental and physical ailments: the complaints that patients bring to their doctors do not fall neatly into two separate packages, one calling for medical intervention or therapy, the other calling for moral counsel and spiritual advice. Rather, the doctor is faced with human individuals whose afflictions embody complex patterns of mental and physical disorders. The practical doctor whose mind is addressed to the real-life character of a patient's complaint will therefore think of his 'physiological mechanic' role and his 'psychotherapist' role simply as ideal types, and so as alternative hats that he can don or doff as occasion requires. And, if he is prepared to help the patient to the limits of his capacity, he will also understand that the 'lawyer' role, the 'priest' role, and the 'welfare officer' role are other ideal types, which he will be justified in taking on – at least on an elementary and temporary level – if only in the hope of changing the patient's human condition to one within which medical or psychotherapeutic remedies can hope to show their own characteristic virtues.

What is there then in common to the professional roles of the physician, the psychotherapist, the priest, the lawyer, and the rest? All these professionals are there to stand by our sides, and help us to find our ways through dark places. From the standpoint of the average human sufferer, the prospect of confronting the majesty and mystery of the law-courts, with life, liberty, and/or fortune at stake, is an intimidating one, which he cannot normally be expected to undertake single-handed, without professional help. The onset of serious disease, or the experience of neurotic anguish, similarly plunges the individual into a 'dark place,' in which he needs help and counsel beyond anything he can do for himself. The approach of death, likewise, is an experience through which, in earlier times at least, most people could expect the help and guidance of minister or priest. When the poet Dante visited the *Inferno*, even he needed the help of Virgil to help him find his way back out. Some of our troubles, of course, may be beyond remedy; many of them will call for management, rather than for 'cure;' but the functions of the professional do not cease at the point at which he exhausts his power to remove completely the original grounds of our complaint.

Where, then, do the concepts of mental health and mental illness fit into this broader spectrum of afflictions and treatments, complaints and remedies? In point of *practice*, they are hard to separate from the general run of medical afflictions and complaints; and any attempt to insist on differentiating them in point of *theory* is liable to fall into the same Cartesian traps that are embodied in the mechanistic view of somatic medicine that is rightly under criticism in this debate. In point of *history*, modern psychotherapeutic treatment became differentiated from the rest of medical treatment at the moment when Joseph Breuer realized that the function of listening to a patient's complaints was not merely to collect symptoms to serve as clues to some underlying bodily defect, but that the best way for him to help Anna O. might simply be *to let her go on talking.*

Finally, in point of *method*, the wisdom of psychotherapy lies in the recognition that the remedies that it provides do – after all – have their own special 'virtues': that the doctor can rightly be just as pleased to see a patient leave the consulting room with mental confusions resolved, with conflicted intentions sorted out, with a better sense of personal direction and enhanced self-knowledge, as to see him leave with an infection cured, with his blood pressure lowered, with a heart condition under effective control, or with a fractured tibia cleanly rejoined. And, if he leaves the doctor with *both* kinds of remedies provided, as well as with a letter of introduction to a trustworthy lawyer or the director of the local social security office, so much the better. 'After all,' the doctor may reflect as the door closes, 'that's just what I'm here for!'

University of Chicago,
Chicago, Illinois

NOTE

[1] Stephen Toulmin, 'Concepts of Function and Mechanism in Medicine and Medical Science,' in *Evaluation and Explanation in the Biomedical Sciences*, edited by H. T. Engelhardt, Jr. and S. F. Spicker (Dordrecht, Holland: Reidel, 1974), pp. 51–66.

D. L. CRESON

MODELS AND MENTAL ILLNESS

The papers of Stephen Toulmin and Alan Donagan raise questions about the nature of those human states that in Western societies have traditionally been called mental illness. Their remarks are directed principally at the basis of our understanding of those states, and the results of that understanding in the manner in which we relate to and deal with individuals who are troubled in their personal lives. In their logical arguments, which are based on assumptions, both authors broach the perennial human dilemma of reality, what is 'real' as compared to what seems real, or what is social or personal reality, as opposed to phenomena that we can in some degree depend upon outside the perceptual and cognitive boundaries of a sociocultural system. I shall refer to Donagan's paper first.

Alan Donagan sketches the history of the shared social and scientific assumption that both 'bodily disease' and 'mental disease' are deviations in the functioning of homeostatic systems. Freudian theory, as Donagan defines it, with its dynamic interplay within a tripartite typology composed of the id, ego, and superego, shares with current theories of physiology a reliance on the homeostatic principal. Understood in the Freudian framework, the concept of 'mental disease' is at odds with the 'traditional conception of man as an autonomous rational animal', and Donagan proposes certain modifications to bring the theory of Freud and theories similar to that advanced by Freud and his followers, into congruence with the 'traditional view' of man.

Professor Donagan bridges the theoretical disparity between the traditional view of man and the Freudian view by suggesting that mental illness be identified with loss of control in those activities normally considered volitional, an inevitable consequence of human existence as we know it. In making such a recommendation, Donagan appears to suggest that mental illness is the norm, in the sense that such a loss of control occurs at times to most if not all people, a paradoxical theoretical position. Despite this apparent problem, Donagan's position is supported by an increasing number of psychotherapists. When Donagan suggests that Freud was mistaken when he likened the ego in its relation to the id to a man on horseback who has to hold in check the superior strength of the horse, he is not alone. Sheldon Kopp has written that '. . . there will be no struggle once he (the individual) recognizes himself as a centaur' ([4], p. 4).

H. T. Engelhardt, Jr. and S. F. Spicker (eds.), Mental Health: Philosophical Perspectives, 71–77.

When the question of reality is broached, however, it is never entirely satisfactory simply to rework old models. If we are to throw out the homeostatic assumptions of the Freudians, what is to be gained by leaving the residual without a close examination to determine what purpose it is to serve? Such a discussion lies beyond the scope of Professor Donagan's paper. He does suggest, however, that mental illness is something very different from the product of transactions within the Freudian typology. If it is so, if neurosis lies incipient in the texture of a model of the world and its inhabitants that one manufactures from the raw material of one's experiences in that world, and if one 'cures' neuroses by experiencing the world or some aspect of it in a different way, then perhaps we would do well to look elsewhere for a theoretical model to aid us in understanding and dealing with behavior we perceive as evidence of mental illness. It is possible that the model of the world that individuals reference in their behavior is not that different in its nature from the models we call scientific and utilize to give meaning of a different sort to that same behavior.

Stephen Toulmin's paper notes the important role of Thomas Szasz and the book, *The Myth of Mental Illness* [8], in what is not a ubiquitous preoccupation with the nature of human behaviors that are labeled as evidence of mental illness. Much of this preoccupation, it is true, appears redundant and unproductive. However, before we dismiss the 'point at issue' as simply 'rhetorical' and therefore of little practical concern, we would be well advised to look carefully at the significance of the preoccupation itself. Perhaps what is most important in the debate is not what is being said about the existence or non-existence of 'mental illness,' but rather that its existence is being questioned and on such a wide scale.

Stephen Toulmin, like Alan Donagan raises issues about the nature of mental illness. He points to the derivations of the term 'illness' and its historical meaning, which is more akin to 'complaint' than to 'disease.' Unlike Donagan he finds a common basis for understanding physical and mental illness in the discomfort that they both produce in individuals who are seeking solace and assistance from physicians.

Toulmin's analysis of the casual and simplistic manner in which the term 'medical model' has been used and abused in the ongoing debate on the nature of what is called 'mental illness,' and his development of a case for the term 'mechanistic model' is useful in conceptualizing the issues in that debate. However, when he suggests that illness as it relates to medicine references 'human affliction, suffering, and pain – in short, . . . the whole range of human "ills" . . .' ([9], p. 66), a number of practical problems become apparent. It is true that with such a construct we can lay aside the troublesome idea of a 'cure,' but only by creating new problems. In such an event the social role of the physician becomes almost messianic in its breadth and the scope of the medical mandate dramatically changes.

Both Donagan's 'homeostatic principle' and Toulmin's 'mechanistic model' have proved valuable in understanding and developing technologies to deal with somatic disease and injury. There is no evidence to suggest that that usefulness has been exhausted. It may well be that it is potentially just as dangerous to lump physical illness with mental illness as the reverse, the equating of mental illness with physical illness.

Perhaps I have over-interpreted Professor Toulmin's position. I expect that is the case. As a practical matter the idea of a humane physician who sees those who come to him as much more than a broken machine is, at least to me, highly effective. There is some evidence that medical educators are just beginning to realize that such attitudes do not just happen but must be promoted at every level of instruction. I do hope, however, that the physician who sets my broken leg is able to conceptualize the fracture in mechanistic terms while dealing with my pain and anxiety in an empathetic and understanding manner.

The problem both speakers contend with is the sociocultural view that assumes some commonality between what is understood as somatic illness and what is understood as mental illness. Professor Toulmin has noted that the social conception of disease changes through time and from one sociocultural system to another. It may well be that the current debate on the nature of 'mental illness' is important as evidence of a contemporary shifting of the social point of view.

While the authors, Donagan more explicitly, imply that the behavior we interpret as evidence of mental illness and the pain experienced by 'afflicted' individuals is akin to normal behavior and ordinary human suffering, both seem to overlook some of the consequences of such a view.

Perhaps Gregory Bateson is right when he says that 'All science is an attempt to cover with explanatory devices – and thereby to obscure – the vast darkness of the subject' ([1], p. 280). Science is not unique in its insistence on a set of assumptions or a model to obscure the darkness by explaining and giving meaning to human behavior and the observable context in which that behavior occurs. Each human entity develops from its inception a model of the world in which it exists and of which it is a part, or, in Goffman's words, it develops 'frameworks' that define relationships and thus provide meaning for experience [3]. Like scientific models, these individual models are not static but under constant pressure to change in response to new data; and like scientific models they resist change. The meaning of the data is made to conform to the model rather than the model to the data. It is an effective model when it provides continuity to human experience and to scientific inquiry. Scientific models are unique only in terms of the rigor of the criteria that must be met if they are to be accepted as 'scientific truth' rather than 'ordinary horse sense.' Robert Pirsig's brief description of a boy who asks his father if ghosts are real is particularly helpful in this regard. The father answers that ghosts are not real in that they are unscientific. When pressed by the boy the father adds: 'They contain no matter . . . and have no energy and therefore, according to the laws of science, do not exist except in people's minds.' After a pause the father continues: 'Of course . . . the laws of science contain no matter and have no energy either and therefore do not exist except in people's minds' ([7], p. 30).

Dr. Ronald Leifer has a much better gift for words than I. He has written:

> Health and disease are neither structures nor functions of the body as are, for instance, the heart and the circulation. They cannot be discovered and investigated in the same way as can a nerve tract or a hormone. 'Health' and 'Disease' are labels used to classify and denote bodily states in medicine, but they themselves are not bodily states. Diseases, unlike material objects, do not exist independently of the purpose and designs of scientists. Instead of asking 'what are "health" and "disease"?' therefore, it is more appropriate to inquire into the situation and purpose for which these terms are used ([7], p. 58).

In essence 'health' and 'disease,' like ghosts, contain no matter and have no energy; they do not exist except in people's minds. In such a case, as Toulmin has suggested, 'illness' or even 'disease' may be redefined in broader terms than is currently the case, or their origin and therefore their meaning may be conceptualized differently, as Donagan recommends. Such concepts, however, are 'real' only if they are believed.

Whether 'complaint' is synonymous with illness or whether 'neurosis' is a loss of voluntary control poses us no problem as long as we believe one or the other. Both are plausible, both useful. Once believed they determine our behavior in response to the behavior in others we believe they explain. In this sense they do not differ from the individual and personal model of the world that acts as a reference for the behavior of those said to be afflicted with a mental illness.

As Thomas Kuhn describes the evolution of science, it proceeds through the quantitative accumulation of data until some critical point when there is a 'non-cumulative developmental episode in which an older paradigm is preplaced in whole or in part by an incompatible new one' ([5], p. 91).

Other social paradigms or models may change more slowly but the 'scientific revolutions' described by Kuhn have parallels in sudden and dramatic social and individual redefinition of 'reality.' It is to this possibility that I referred earlier when I mentioned the possible significance of the present debate on the nature of 'mental illness.' Both Professors Toulmin and Donagan are participating in that debate. Their papers reflect the more thoughtful aspects of the controversy. Such scholarly work is paralleled by polemical exchanges in professional as well as lay sectors of the sociocultural system.

Thomas Szasz's provocative book raised the issue of the nature of mental illness, but the issue already existed. It was explicit in the early work of Gregory Bateson, Erving Goffman, and a small but scholarly group of systems theorists and communication researchers, and implicit in many of the psychotherapeutic technologies, such as gestalt therapy and transactional analysis, that were being developed at that time.

Both Toulmin and Donagan suggest that individual human entities will continue to be confused and conflicted whether we call them ill, diseased, neurotic, or simply troubled. Although neither author specifically discusses behaviors such as so-called hallucinations and delusions, the hallmark of what are now considered the more severe types of mental illness, I feel confident both would agree that such manifestations are not simply going to disappear. That means we must explain them and the meaning we give them must be compatible with our view of the world, our model of reality. That is what both Professors Toulmin's and Donagan's papers are about. They both suggest changes in our model for giving meaning to suffering and disconcerting behavior.

I want to suggest another perspective. We can use these authors' struggle to conceptualize mental illness in a more useful way, as a model for understanding and dealing with the psychiatric patient. Perhaps as Donagan implies, troubled and conflicted individuals change their view of their world as the result of new data based on new experience, in the same way that 'scientific revolutions' occur when old paradigms, despite the best efforts of scientists, are inoperable in the face of the data. In such a case the humane physician described by Toulmin should be able to develop a new and more efficient technology for dealing with the anguish of his patients.

Various theorists and clinicians are already working to develop such technologies. In their recent book, *Change*, Paul Watzlowick, John Weakland, and Richard Fisch [10] report on a psychotherapeutic approach they have developed for rapidly modifying human behavior that would traditionally have been considered pathological. They use the inevitable paradoxes that are implicit in the nature of models or paradigms whether they are scientific or personal. When the implicit paradoxes in a scientific model become manifest there is a 'scientific revolution.' The psychotherapeutic approach of Watzlowick, Weakland, and Fisch is designed to bring about change in an individual or group in exactly the same way, by making manifest the implicit paradox in the individual's view of his world in his attempts to carry on his daily activities.

Earlier I stated that both Stephen Toulmin and Alan Donagan broach the human dilemma of 'reality.' They do so by questioning existing paradigms in medical theory. I do not know what the 'reality' of tomorrow will be with regard to human behavior but I am certain it will be different from the 'reality' of today. I am, however, grateful to scholars like Professor Toulmin and Professor Donagan who through their presentations bring to the fore some of the paradoxes implicit in my own assumptions about human behavior and 'mental illness.' Change as well as constancy are, after all, characteristic of each of us as individuals even as they are of science. In Paul Watzlowick's words: '... contrary to general belief, order and chaos are not objective truths, but – like so many other things in life – determined by the perspective of the observer' ([11], p. 57). To bring 'order' to 'chaos' may be the most basic of human drives, for it is only in 'order' that we may function productively. The so-called mentally ill individual has imposed the appearance of 'order' on his world, as we all do. The order he has imposed, however, is at the expense of personally rewarding and socially approved behavior. His problem may not be chaotic thinking, but the nature of the personal order to which he refers in his behavior. Such an understanding of psychosis is just as applicable to the prevalent human conflicts that characterize the "normal" daily activity of human life.

University of Texas Medical Branch,
Galveston, Texas.

BIBLIOGRAPHY

1. Bateson, Gregory: 1958, *Naven*, Stanford University Press, Palo Alto.
2. Donagan, Alan: 1977, 'How Much Neurosis Should We Bear?', in this volume, pp. 41–53.
3. Goffman, Erving: 1974, *Frame Analysis*, Harvard University Press, Cambridge.
4. Kopp, Sheldon: 1972, *If You Meet the Buddha on the Road, Kill Him!*, Science and Behavior Books, Inc., Palo Alto.
5. Kuhn, Thomas: 1962, *The Structure of Scientific Revolutions*, University of Chicago Press, Chicago.

6. Leifer, Ronald: 1969, *In the Name of Mental Health,* Science House, New York.
7. Pirsig, Robert: 1975, *Zen and the Art of Motorcycle Maintenance,* Bantam, New York.
8. Szasz, Thomas: 1961, *The Myth of Mental Illness,* Hoeber-Harper, New York.
9. Toulmin, Stephen: 1977, 'Psychic Health, Mental Clarity, Self-knowledge and Other Virtues', in this volume, pp. 55–70.
10. Watzlowick, Paul, Weakland, John and Fisch, Richard: 1974, *Change,* W. W. Norton, New York.
11. Watzlowick, Paul: 1976, *How Real is Real,* Random House, New York.

HORACIO FABREGA, JR.

DISEASE VIEWED AS A SYMBOLIC CATEGORY

I. INTRODUCTION

The essays in this volume address some of the social and philosophical problems which surround the phenomenon of psychiatric disease in contemporary society. This is all being handled by reference to notions such as 'mental health' or 'mental illness,' terms which I tend not to use. Nonetheless, the problems are quite 'real,' have caused me concern, and have been influential in motivating my academic work. Not surprisingly, as a physician I have concentrated my work on the meaning of 'disease.' I have tried to develop a general frame of reference about disease. Interested in the beliefs, orientations and actions of various people – people from cultures far different from our own – I have studied and compared how disease is viewed and handled. In this paper I would like to summarize some of my thinking on this topic. An underlying theme in my work has been that, by comparing how different people think of disease, we will be able to develop a broad perspective about what disease means and the role it plays and has played in the lives of different people. Such a perspective may provide us with clues about the sources of some of the problems involving disease and medical care which exist in our own society.

At the outset, let me say that something similar to what the average person refers to by the term disease or illness in our culture is found in all societies which have been studied carefully. All people, it seems, set apart a large class of undesirable events and changes which can affect them in a personal way; a subclass of this all embracing one is referred to by linguistic terms roughly analogous in meaning to what we mean by disease, illness, sickness, etc . . . The phenomena referred to by them through their vocabulary of disease and illness is similar to the phenomena we label by means of our own. Some of the differences which one notes and its implications will be discussed later. For now let me say, for purposes of discussion, that the beliefs, orientations and actions which can be linked to this category 'illness' or 'disease' may be viewed as the medical domain of a cultural group.

Given this generalization about the ubiquity of disease and the fact that almost by definition it is associated with changes which are undesirable,

H. T. Engelhardt, Jr. and S. F. Spicker (eds.), Mental Health: Philosophical Perspectives, 79–106.

people have articlated beliefs about why and how disease is produced and also how its burdens may be counteracted. The beliefs and understandings of disease of non-literate people are ordinarily studied from the standpoint of how correct or useful they are in the light of our own biomedical orientation which we assume is factual or correct. My orientation has been different. I have tried to approach disease largely in terms of the cultural beliefs and orientations which guide people to various forms of 'medical' action. This emphasis on the cultural medical perspectives of different people is what I mean by the title of my paper, 'disease as a symbolic category.' If one approaches the study of disease in this way, then all types of beliefs or theories of disease are relevant, regardless of how developed or simple one may judge the group to be. This is so because the phenomenon, disease or illness, is viewed in terms of the group's system of signification. This, in a sense, means that there are several forms of disease, not just our biomedical one. The purpose of this paper is to discuss medical beliefs and orientations from such a frame of reference. In doing so, emphasis will be given to generic features of these beliefs and an inductive comparative method will be employed. The paper attempts to provide a synthesis and structural account of beliefs about disease, using as data information about the medical beliefs of non-literate and modern groups. This article will also briefly touch on the relevance of these beliefs for the understanding of behavior in the event of disease and for understanding the contemporary disciplines of medical in general and psychiatry in particular are briefly touched on as well.

II. DISEASE AND THE QUESTION OF PERSONHOOD

Although it may be true that an actual occurrence of illness or disease in any person, regardless of his culture, is associated with changes in his chemical and physiological systems, it needs to be emphasized that this is now *our biomedical interpretation* of the matter. Biomedicine constitutes an evolved and highly specialized view of disease. For purposes of analysis, biomedicine should be viewed as a *medical perspective* and not as the correct answer about disease. The impact of biomedicine as a social institution and system of medicine goes back but a few centuries. When seen against the span of the history of man as a biological species, biomedicine occupies a very small segment indeed in what is a very long line of human activities. Contemporary non-literate people, and no doubt people of earlier prehistoric epochs, do not look at disease in terms of enzymes and X-ray shadows. Rather, disease appears and is 'seen,' so to speak, as a change in the behavior of the individual;

a change which involves verbal reports of various types linked to an impairment in the individual's ability to carry on socially as a member of his group. This seems to me to be a basic ingredient of the generic disease: it involves an interference in the individual's capacity to carry on his or her actual or expected social activities, and it is in the social sphere where it shows itself and has its effects.

A central feature of disease, as this is experienced by the individual, is that it brings about or forces a change in his definition of himself. Whatever language symbols an individual draws on from his culture, in order to define himself as a competent being, this image of himself is altered during disease. There is an incongruity between an ideal image of self and the new prevailing experience of self. Because of this one can say that disease involves an alteration in the way an individual is symbolically characterized, either by himself or significant others. A person who is sick has thus undergone a transformation into a negative state. In this light, the pursuit of medical care or treatment can be viewed as a search for a transformation back to an ordinary and relatively 'normal' status or identity, although one may hold that permanent changes have occurred. These conditions related to disease take place regardless of whether one observes a sick person in a so-called 'primitive' culture (who believes himself or is judged by others to be punished by the gods or maligned by his enemies) or one in a modern industrial nation who believes that his chemistries are altered.

I am of course viewing things here very generally. Characteristics about disease which are intuitively compelling have been disregarded. Thus, I am playing down what some may view as the 'obvious' fact that disease involves a disorder in the physical body, or an altered biological state. This has been for two reasons. The first is that the term 'body,' as ordinarily used, is tied to our own Western philosophic premises and draws part of its meaning from the contrastive category 'non-body,' ordinarily phrased as 'mind.' I am here handling dualism, and the reductionism that goes with it, as a cultural perspective. In our culture and civilization this perspective, and its application to medical problems, has been enormously useful. It has channeled an approach to disease which has allowed us a great deal of control. At the same time, however, this particular perspective has beclouded many general aspects about disease and illness. To non-literates and non-Western people (and probably to 'man' seen in an evolutionary time frame), dualism and the idea of an impersonal and mechanically functioning 'body' in the context of disease has little meaning since the changes which are implicated in disease are judged quite differently. The person and agencies outside of him are changed during disease; and disease results from natural and preternatural forces which are held to directly connect with the person ([4], 23). Although one can say that all persons have bodies and minds, or alternatively, that persons' descriptions of themselves can be read as describing either mind centered or bodily centered changes, it seems prudent for now not to rely on the Western dualistic approach which tends towards reductionism.

Rather, I shall continue to speak of a generic disease and view it as simply involving a set of complex and culturally contextualized *experiences*. These are signified by linguistic terms which are formally and semantically analogous to our 'disease,' 'illness' or 'sickness.' The experiences certainly include such matters as pain weakness, malaise, and perceptions indicating disordered vital functions. Moreover, an observer might link these causally to altered processes in the 'body'. However, because my goal here is to develop a broad and comparative frame of reference the perspectives of non-Western people need to be preserved. This is the reason for avoiding both the dualistic Western view of a separation of mind and body and the reductionistic somatic view which gives a depersonalized portrayal of the human body.

There is a related reason for avoiding dualism and its associated reductionistic perspective in a general symbolistic formulation of disease. It stems from the observation that persons (regardless of their culture) do not really bring disordered bodies or minds to a medical practitioner. Persons bring, instead, aches, changing images of self, worries, concerns, altered perceptions, and experienced constraints (Dr. Toulmin has developed this point beautifully in his paper in this conference'). Disease can thus be viewed as involving, or 'made up,' of just such types of undesirable sorts of changes. To the individual his 'body' is not the impersonal and mechanical entity that it is to a scientific observer. Alterations which we can specify in biomedical terms give rise to a number of private experiences [11]. In these personalized representations of an altered self – which embrace our notions of body and mind – one finds an important feature of culture, considered now as a system of symbols which the person literally uses in order to explain how and why he functions, how and why he becomes diseased, and also who he is and what his world is like.

Parenthetically, one can add that, if one had a frame of reference or metalanguage, which accommodated however coarsely, diverse representations of self, one would be able to study and compare how disease is construed and how symbols of disease function phenomenologically among various people [10]. Symbolic systems, involving the self and ways in which the self can be altered, serve to partially guide and model social behavior during instances of disease. And behavior, in turn, is given significance in terms of just such types of symbols which are part of the culture of a people. In this sense, symbols about disease, which underlie behavior, pattern and shape medical treatment. This link with behavior underscores an additional fruitful consequence of studying disease symbolistically. Comparative studies of medicine are very much in need of language of discourse which addresses the social implications which disease has in the group and emphases on behavior can accomplish this [5]. The study of the symbolic and behavioral correlates of disease, then is very likely to contribute to the eventual development of an ethnomedical science and to an ethnomedical theory of disease.

III. THE NOTION OF A TAXONOMY OF DISEASE

As one surveys the literature in anthropology which involves medical problems, and the history of medicine as well, the following generalizations can be made: because occurrences of disease are disruptive and undesirable to the person and his immediate social group, they give rise to questions about existential, moral and physical placement. If persons were perfectly functioning entities – if there did not exist altered states that we term disease – then people and groups might be less driven to inquire into who they are, why they are the way they are, and why they function the way they do. Certainly many other kinds of undesirable changes would remain as realities and possibilities, and with it the sense of one's basic vulnerability and powerlessness. However, formulations about the world and about man himself would be quite different were he deprived of the personalized experiences surrounding disease.

This view of matters has led me to judge a group's beliefs about and classifications of disease as devices which allow the individual to construct explanations or formulas about how and why he has become changed in a negative way. These formulas are derived from what I have come to term the group's *taxonomy* of disease. In general anthropology and biology, a taxonomy is described as a group's way of classifying, ordering, and naming a domain of interest – say the domain of animals, or plants, or types of firewood, or kin types etc . . . . A taxonomy also reflects (and may be taken to logically imply) a *theory* which explains or gives a rationale of how that domain 'worked' or is organized. In naming and explaining an area of relevance to a cultural group folk taxonomies provide a way for members to relate and behave in a culturally appropriate manner to that area of their world.

In this light, a group's *medical* taxonomy may be viewed as an ordering scheme which classifies and names diseases and provides explanations about how and why they occur. The changes in personhood bound together in disease are rendered coherent by the taxonomy. Now, since disease is by definition undesirable and unwanted – people in general want to be rid of disease – medical taxonomies provide the individual with a way of handling his disease so as to neutralize or eliminate it. In short, if one posits that disease is a negative and undesirable condition of an individual, then taxonomies provide explanations for this condition and also rationales for treatment. The range of possible meanings that a condition of disease can have are also captured by this taxonomy.

The medical taxonomies of non-literate people are usually couched in what anthropologists describe as a supernatural frame of reference [23]. Those dieties, spirits, powers, entities, beings or processes which are believed to influence and sustain worldly affairs are the ones that, in an ultimate sense, 'send' disease, especially those that are judged most troublesome by co-members. In short, quarrels, antagonisms, the breaking of taboos, the essences or products of animals, etc. . . . bring disease. We, of course, explain disease by positing that man is like a machine whose functioning is upset through injury or damage of vital parts. The injury stems from the way the machine was constructed (e.g., genetic predisposition or 'defects') or from environmental 'irritants' of various types. Here, in this taxonomy, natural and not supernatural factors explain disease. This distinguishes the general meaning of our taxonomy of disease from that of non-literates. However, a basic point of agreement is that although diseases affect persons, their ultimate bases or sources are somehow outside the individual. What one could view as a basic problem in accounting for this transfer of effects and their associated changes (i.e., explaining how disease comes to reach and affect *concrete* persons), can be used heuristically to interpret and make sense of taxonomies of disease.

IV. DOMAINS OF MEDICAL TAXONOMIES: THE REGION OF THE PERSON

One can differentiate three broad regions which to varying degrees are implicated in explanations of disease and directives for treatment: the region of the *person*, the region comprising the *worldly environment* (social and physical) of the person, and the region comprising the *other worldly* or *supernatural* environment of the person.

With regard to the first region, processes or forces eventuating in disease, whatever their nature or source, must ultimately establish contact with and then transform the person for it is he who becomes ill. Since such processes alter the person in complex ways this means that to varying degrees, medical explanations may elaborate on just how the person is constituted. In a sense, cultural groups have some option as to how to interpret the way the person 'works.' This point was mentioned earlier in terms of symbolizations of self and personhood. In general, an explanation of an occurrence of disease will involve a consideration of different parts of the individual and how these parts interact. Quite obviously, one can anticipate that the taxonomies of different people will differ greatly when one focuses on this particular region.

As an example of this point, a 'disease' may be believed to enter the individual during sleep (i.e., a spirit, a wind or punishment) with the portal of entry not differentiated specifically but metaphorically – e.g., sleep entails an opening to, a vulnerability towards, or a state of readiness to communicate with preternatural and/or potentially harmful influences. Furthermore such influences may be believed to simply bring out global and poorly differentiated changes in the person. In other words, the articulated features of disease – what we refer to as the signs and symptoms, functional and structural changes, etc., may not form part of the explanation of an occurrence of disease. Somehow, the whole disease is merely a consequence of a harmful influence which entered or touched the individual, but *how* the individual is changed during disease is not considered important in individuating the disease and specifying a rationale for treatment. This would seem to constitute a simple and straightforward way of representing the region of the person during disease.

In contrast to this would be a representation which includes the entering of the 'malady' (e.g., either as a germ, or poison, or spirit, or evilness, or magical influence, etc. . . .) in the person's food (portal of entry being the mouth, with ingestion a key process). The person may be believed to have been punished by specific types of spirits because of discrete transgressions, perhaps as in the above instance. But the consequence of these matters may have been a tightening or constriction 'of the stomach' (e.g., in the case of prominent abdominal symptoms), an excessive warming and hardening of the blood (in the case of localized pain), and the 'capturing' of certain vital parts within the apparatus which render the individual weak (in the case of vague symptoms of malaise and weakness). In short, the concrete aspects of the person – namely, what we in Western culture term his 'body' – may also be incorporated in the explanation of an occurrence of disease, and this explanation involves putative regions, mechanisms, and functions (however crudely or accurately these explanations might appear when judged from the standpoint of Western physiologic ideas).

It should be re-emphasized that I view medical taxonomies not only as classifying disease types or providing causes or simply names for them; but, as also providing a rationale and an explanation of disease, and a plan for treatment. So called 'explanations' derived from taxonomies of disease accomplish all of these things. Moreover (and I believe that this has important implications for our theme in the conference), taxonomies are internalized by the individual and thus influence his thinking and feelings during disease. Taxonomies of disease, cultural traits of a group, thus contextualize disease for members of that group. The many changes taking place in the individual during disease – changes reflected in what the person reports and does – are rendered meaningful through the individual's taxonomy of disease. For this reason one can say that much of the behavior of the person who is sick is regulated by his taxonomy of disease. In this light, if a taxonomy incorporates in a rich way the region of the person, then it would seem that his experiences and behaviors during disease are somehow more drawn to his own being (i.e., his body or his mind). In this instance, we, who endorse the biomedical taxonomy of disease, would say that the psychophysiologic correlates of disease are more likely to be individuated during the state of disease. Actions of the individual during disease naturally come to be explained in the group by premises which anchor the taxonomy of disease.

I must reiterate that the accuracy or correctness of the way the group divides up the region of the person in explaining disease is not important in this phase of an analysis of a group's system of medicine. What is important is simply how and also the degree to which the region is specified and articulated as a phenomenal structure. The implicit assumption is that social groups have to relate to a relatively complex natural world which includes the concrete person. This natural world lends itself to degrees of differentiation, and codification. It is the individual, by means of his culture, who literally partitions the world and specifies its attributes along varying dimensions, though regularities in the world, to be sure, may themselves serve to pattern and set boundaries on the processes of perception and cognition [1, 3, 12, 15]. In the personage of a concrete individual, cultures are provided with a complex and enigmatic entity; and, in the changes that are disease which span across the segments or parts of the individual and which have to be transcribed and rationalized by the person cultures will 'demonstrate' the richness of their medical taxonomies.

V. DOMAINS OF MEDICAL TAXONOMIES: THE REGION OF THE WORLDLY ENVIRONMENT

The same reasoning may be used to analyze the region of the worldly surroundings of the person – the physical and social environment. Regardless of where disease is held to come from, how it is judged to be produced, and how it is held to change the person the influences culminating in disease eventually have to cross the worldly environment since it is there that man is placed. Among non-literates, the communication between what may be, on the one hand, supernatural influences, and on the other, the actor proper, may of course be direct – as in the first example provided above. In many (perhaps most) instances, however, the presumed way in which such influences establish contact with the person involve more elaborate connections in the worldly environment. Of course, the sources of disease may reside entirely in the environment itself [16].

As an example, what were termed supernatural influences may be believed to be situated inside mountains or in abodes above the heavens. Disease may be judged as an outcome or proximity to the mountains or due to the commission of certain classes of action within specific territories of the physical habitat which, as it were, angers the gods or spirits. Viewed symbolistically, then, the various territories of the environment might be types of connecting bridges that link preternatural influences with concrete persons. Certain types of trees or emanations from rivers or lakes may be held responsible for disease. Alternatively the influences may be seen as further away – e.g., in the sun, communicated as heat, or closer, as involving nearby persons, animals or their exuviae, or contaminated or 'spoiled' foodstuffs, or even germs or bacteria which are somehow inhaled

In short, some of the influences which constitute the sources of disease (whether we may view them as 'real' or 'imagined') must be presumed to be connected to or located in the physical habitat: to have a tangible form, to be visible or invisible to the naked eye, to be natural products or artifically created ones, etc. . . . In principle it should be possible to construct a symbolic calculus or classification of types of environmental entities and of their ways of connecting with the person. Rules or formulas probably underlie how worldly entities become implicated in disease; such rules remain to be discovered and when they are some of them may be shown to have wide applicability across cultural groupings.

Obviously if the *social environment* is singled out, which is to say that social actions or relations are implicated in the causation of disease, then one must be prepared to have available a calculus or classification of distinctive types of persons, relationships, and conflicts with which to analyze taxonomies of disease. Such classifications and the rules for deriving how they produce disease will have to be worked out in order to comparatively analyze medical taxonomies.

When one accepts the premise that the entities and process of both the physical and social (worldly) environment of man can be differentially implicated in disease, one is entitled to ask the following question: in a given explanation of disease which is derived from the group's taxonomy, which of the many types of ecological elements are implicated, at what level of abstraction, and how richly and systematically are they linked with the concrete actor? Indeed, by initially weighing the extent to which one or the other of these worldly elements (i.e., social vs. physical) is implicated in disease, an important analytic requirement would be accomplished. A review of the literature in anthropology will disclose any number of ways in which the worldly region of the person – his physical and social environment – is implicated in disease, and each of these involves different degrees of differentation. As in the discussion above, one may disregard the notion of the validity of the explanation – the so-called 'scientific' truthfulness of either its premises and/or key concepts. We can focus instead on its properties as a symbolic system, on the degree of specificity and differentiation of the symbols, and on the degree of intricateness in the way they are used to produce appropriate explanations of disease – with explanation, again involving name or type, mode of causation and rationales for treatment. Alternatively, one would be searching for the degree to which the taxonomy articulates a differentiated environment, and the complexity of the rules which explain how entities work and interact to produce diseases.

VI. DOMAINS OF MEDICAL TAXONOMIES: THE OTHER WORLDLY ENVIRONMENT

Clearly, the preceding approach to the study of taxonomies of disease could be applied to the third of the regions that can be implicated in an occurrence of disease, namely that termed 'other worldly.' Turner has cogently remarked that the examination of this region involves dealing with supernatural phenomena: material, entities, powers and the presumed modes of action and aims of phenomena seen *other than as natural* [23]. This 'region,' which is often implicated in explanations of disease, may have to be conceptualized independently from the worldly one, since groups, in explaining and dealing with disease make this distinction and communicate, i.e., draw correspondences, between them. Such things as spirits, beings, forces, and dieties may be modeled in terms of worldly or natural things or occurrences (see above). However, they may also need to be judged as symbols whose referents stand outside the natural world, here often referring to other agencies and powers which control the worldly setting. Again in theory, it should be possible to enumerate, weigh, and classify such types of preternatural agencies, and apply the principles of analysis discussed earlier, i.e., by degree of articulation, differentiation and systematization. The sheer number of classes of spirits and demons which can cause disease can be expected to differ, or groups may have hierarchies of levels of them that differ. The actions of these may stem from complex deliberations which they have among themselves. There may, in fact, be limits to the ways in which these preternatural agents can produce disease. It may be possible to render by way of formulas or logical calculi what these limits are such that this region of medical taxonomies can be weighed or quantified and compared across people [5]. Clearly, the feeling states and the experiences of the individual who is sick may reflect his conviction of being manipulated and controlled by such supernatural agents, so that analyses of these aspects of medical taxonomies will clarify the behaviors and experiences of persons who are sick. Supernatural agents are often held to control and act through physical entities in the environment. For this reason, treatment efforts require that rituals be performed in the physical world, though the incantations and gestures which make up the ritual are aimed at the world beyond. This region of the taxonomy, then, bears an obvious relation to the study of ritual. To the extent that the preceding kinds of emphases prevail during disease and medical treatment, they would seem to involve an instance when the supernatural region was more heavily represented in the explanation of disease.

To summarize, one can say that taxonomies of disease offer members of groups 'rational' ways of explaining and dealing with disease. The unique features of disease – that they are problematic, recurring, unwanted and therefore that they give rise to a need for corrective action – account for the fact that taxonomies of disease will tend to bring into play fundamental complements of the culture. I have indicated elsewhere that if instantiations of specific diseases, identified through an independent language, are studied comparatively, it may be possible to draw empirical generalizations that will enhance one's understanding of symbolic systems, of social aspects of disease, and of human adaptation generally [5]. Taxonomies of disease, in short, reveal a group's symbolizations about elemental matters of both intracultural and universal significance. When they are studied in conjunction with suitable (independent) languages of disease that specify properties and components of (a) disease(s), they may provide a means of addressing matters of central concern to biology as well as social science. Finally, since the medical explanations – and treatment practices – of different social groups seem to draw on cultural representations of the three regions singled out above, I believe that it should be possible to develop a metalanguage or calculus which should allow one to compare and relate – one to the other – different taxonomies of disease and their mode of functioning. I accept the point that in certain ways different taxonomies of disease are incommensurable. Nonetheless I feel that one should be able to measure, quantify – in an information sense – and compare how taxonomies of disease are used. That is, how taxonomies of disease, in rendering explanations of disease and rationales for treatment, function comparatively. This point will not be developed further here, however [7].

VII. PROBLEMS ASSOCIATED WITH THE STUDY OF HOW MEDICAL TAXONOMIES ARE USED

An interesting feature of the way medical taxonomies, compared to other taxonomies, are used by different people, is related to special attributes of disease. Even if one were to adopt the view that disease constitutes a universal naturalistic 'specimen' of sorts which cultures merely 'label,' the fact is that disease does not always appear fully formed. The properties of disease need not be present all at once. To the extent that the explanation of a disease may be influenced by its perceptible properties, then problems are posed since members simply do not see all of these at any one time. Rather, what one observes, is that disease *unfolds*. In other words, diseases have a temporal extension and indeed an occurrence can, on logical grounds, be seen as having different identities in time, as indeed is the case in some groups. In addition to reflecting underlying physiologic changes, this property of disease is also a consequence of the fact that disease resides in individuals who are parties to any number of ongoing social experiences which are constantly being influenced by situational factors. The unique unfolding character of disease also follows from the fact that the many influences which bear on an occurrence (e.g., what we term biological, social, psychological, etc. . . .) are interconnected and feed back upon each other. Finally, since occurrences of disease are, in the last analysis, expressed and grounded in the behaviors of the person who is sick – behaviors which can differ as to visibility – parties to the occurrences can 'read' these behaviors differently. Thus, whereas diseases comprise a dynamic and patterned whole, the objects of other taxonomies involve physical external, and relatively static entities in the world.

As an example, plant specimens are physically enduring entities and so are types of firewood. The color spectrum, likewise, can be seen as fixed and rules for labeling its regions (or core areas – as discussed in [1]) likewise are more or less fixed. Of course, concrete objects may change in color and different color terms may come to be properly applied to an object across time, but the object itself is likely to retain a concrete specificity. Indeed, it is likely that the identity of the object is never really in doubt under these circumstances. With regard to kinship systems, many of their sustaining properties, e.g., characteristics of persons, social relations, generational depth, etc. . . . , are likewise relatively fixed and unambiguous, at least in the short run. In brief, it would seem that other phenomenal domains, that have been studied, share a degree of fixity permanence, wholeness, and visibility. For this reason, taxonomies which classify and explain them may be more easily applied.

A distinguishing feature of disease which sets it apart from referents of other taxonomies, then, is its unfolding character; the referents of disease terms are not all public, given, and fixed but, rather, partially buried, variable and changing. The meanings of disease terms, of course, could be regarded as fixed and to some extent closed (though Frake informs us that they are in some instances open and connected with each other; see [9]), but not the referents to which they are applied Indeed, it is precisely because the referents of disease terms are often ambiguous and can change across time that groups are able to attribute differing and changing significations to disease – they can make it 'be' or 'stand for' different sorts of things.

Given these factors, diagnosis – and other uses to which medical taxonomies are put – can be seen as the process which certifies, sanctions, and/or validates the medical experiences of a people. In explaining a disease occurrence, then, a group reflects the existential meaning which it gives to this (medical) side of its life, and it does so by means of its taxonomy of disease, a device which the group has, as it were, *created.* I would like now to draw attention to a characteristic feature of the taxonomies of disease used by non-literates (and more than likely, of those taxonomies used by earlier pre-historic people). Until contemporary times, people have related to disease principally as social episodes, and have explained these in terms of entities and processes which were interdependent and linked to everyday personal and socially vital concerns. Social happenings, supernatural deliberations, and personal motives, present and past, were linked and indeed formed a part of disease and illness. I express this point of view by saying that disease among non-literates is seen as directly connected to social happenings; it is manifested socially in altered behaviors and in personal constraints and is made sense of by the people in terms of phenomena integral to social relations.

In this light, one can now, perhaps, better appreciate a striking feature of our *biomedical* taxonomy, a taxonomy which makes use of highly intricate parts of the person and which draws on so-called natural and impersonal agents. Because disease is now defined in terms of enzymes and X-rays, and not by considering the individual's social behavior, on logical grounds any episode of illness explained by this taxonomy does not directly draw on or feed into social and psychologic matters. One can say that in this particular taxonomy – the biomedical taxonomy – diseases reflect impersonal and technically malfunctioning sorts of things and are literally socially *nonsensical.* Furthermore, and speaking in general terms, in biomedicine, once a diagnosis is reached, then *regardless* of the differing courses that that disease may pursue, it still tends to have the *same fixed* significance which remains uncoupled logically from social affairs. The person, as it were, can live, get better, die or deteriorate, but he still has the same disease. Insofar as disease in our taxonomy stands separated from social goings on and, furthermore, can be said to be timeless and fixed as to meaning, then disease cannot easily be brought in as a basis for certifying or validating specific aspects of social life that require resolution and certainty. To put the matter succinctly, little of what cancer, tuberculosis, or myocardial infarction portend to persons or to the group can be related to social processes and relations by an appeal to the biomedical taxonomy qua system of meaning.

Among non-literates, however, I have said that things are different; somehow much of what people do socially, think feel, etc., is causally and substantively to their diseases. Moreover, the names and meanings given to disease change during an occurrence. What one might view as an outbreak of nephritis, for example, others may judge food intoxication, then spirit punishment and finally witchcraft. Recall that the idea of a chronic disease, a subclinical disease, and for that matter a converted (i.e., hysterical) disease are biomedical creations. If one looks at this problem of disease naming and specificity comparatively, then interesting questions are raised. The pursuit of these takes one to central questions in philosophy. In this regard, one can ask: What are the grounds for knowing something, or for establishing the correct 'identity' of an entity or regularity in nature? What are the sources for and/or processes through which cultural groups validate experience [19]. Several answers can be surmised; among these, of course, are processes tied to observation, and processes surrounding spiritual revelation. Plant specimens can be named with a measure of reliability since they can be commonly observed, and such observations serve as the boundary conditions for a system of beliefs. Terms which are applied to a particular botanical specimen in space and time owe their (relative) commonality of meaning to directly shared experiences that are focused on relatively unchanging sorts of things in nature which, as it were, persist. Moreover, these 'things' are not inherently problematic. Diseases, on the other hand, come and go, and at what point in time and on what basis is one allowed to claim that 'it' has been observed or 'it' is gone? Put differently, in what manner and on what grounds is a consensus reached about the identity, let alone meaning, of an occurrence of disease? Indeed one is allowed to ask: was consensus, in fact, actually reached? Did the relevant parties to an occurrence of disease in fact *share* information about its identity? If so, on what grounds was such information shared, for which types of occurrences of disease is such information shared, or during which times or in which situations is a measure of consensus reached about disease?

On intuitive grounds, one can say that in many instances it is only *when* a disease occurrence has ended that the person and/or others may *know* what the identity of the disease *was* – the identity of disease is often only 'known' after the fact, as it were. This of course follows from the special character of medical taxonomies; they not only label and name disease, they are also guides to action, and action of a special character. Taxonomies provide the basis for treatment, which by definition is rationally based in the culture and tied to *cause*, and the accuracy of the taxonomic category used is confirmed when the function of the taxonomy has been fulfilled, i.e., the disease has been successfully treated.

In sum, it is often only when treatment has been successful or a patent failure – which then clarifies, if not establishes, the matter of cause – that the identity and meaning of a disease is felt to be understood. This tentative and essentially ambiguous way of certifying elements named and explained through a taxonomy is simply not a striking feature with regard to other types of taxonomies. For this reason, it may be anticipated that special problems will be encountered in the study of medical taxonomies, and furthermore, different functions may be judged as served by the taxonomy. Among non-literates, disease has different significations even in the short run and the taxonomy of disease functions so as to flexibly feed in and out of much of what is important in the social lives of the group. As I have said, in our taxonomy, disease tends to have a signification which is fixed, even in the long run. Furthermore, since its indicators have become separated out of social happenings, disease explaining cannot be easily related to social happenings.

VIII. TAXONOMIES OF DISEASE SEEN AS ELEMENTS OF THE SOCIAL REALITY OF A PEOPLE

In light of earlier discussions, one can appreciate that ascertaining which of the many (folk) diseases which are prevalent in a non-literate group are lethal, becomes a legitimate inquiry in the comparative study of medical taxonomies. Clearly, the identity and explanation of disease need not of *necessity* have a relation to death. In fact, among non-literates, the question of disease identification is often tied *logically* to the issue of its eventual import and outcome. All of this is made possible by factors discussed earlier. These are, the protean, unfolding and problematic character of disease as well as the distinctive way in which disease happens to be defined among non-literates, namely as a social behavioral 'thing' which is connected to human, natural and preternatural phenomena.

In our culture, the concept of disease, the concept of death, and the explanations that obtain about the how and why of the associated happenings tend, by and large, to be logically independent though empirically they are often intertwined. This can be illustrated formally by a consideration of three propositions which in our culture come close to being axiomatic: (a) Death is a possible outcome of *any* disease, (b) Death is a possible outcome of *no* disease, (c) Non-death is a possible outcome of any *disease.* (A counter example here is our class of genetic diseases, members of which, for many reasons, may have to be analyzed differently [5]). The logical independence that exists in our culture between the explanations of death and disease is made consistent with everyday experiences, which indicates that death and disease are often empirically interrelated by means of two associated sets of notions. These are, on the one hand, predications that biomedicine allows, made both about disease (i.e., its severity, intensity, etc. . . .) and the state of the organism (i.e., his degree of vulnerability), and on the other hand, predications that can be offered about the fundamental bases and influences that prevail in the world (i.e., accidents, natural causation, the consequences and limits of human action, etc. . . .).

By means of notions such as these, a measure of logical independence between the explanations of disease and of death is preserved. One is thus able to say 'he has *X* disease,' 'he died from *X* disease,' 'he recovered from *X* disease,' 'he died in an accident' and also 'he was killed' and feel that one is, to some extent, addressing separate classes of happenings and phenomena. One might in fact judge scientism and secularization to be social devices that have brought about or forced this independence between disease and death. Devices, in other words, which generate the view that disease is an inevitable but *socially non-sensical* outcome of man's place in the environment, and that death and finitude are factors which have no underlying control or design.

Among non-literates, it is not necessarily the case that the truthfulness of propositions (a), (b), and (c) is maintained. Death, disease, and associated explanations interrelate logically: both are grounded in social experiences, and the whole is seen to be under *personalized* design or control. Only certain diseases can bring death, whereas others *never* do; and that which lies behind diseases which kill is *the same* as that which may lie behind all death and indeed all problematic human and social happenings. In short, the fundamental causes and bases of disease (supernatural beings, other worldly influences, human malevolence) are usually the same in principle and kind – indeed, equivalent – to those assumed to underlie most deaths that take place (possible exceptions are early infant deaths and deaths of the elderly). In this sense, disease and death are expressions of the underlying order and control felt to exist in the world, and which order is personalized.

In the light of these considerations, it is appropriate to hypothesize that important relations obtain between a group's explanations about the causes of disease and its explanations about the causes of death. Indeed, from one limited standpoint, so called 'primitive' taxonomies of disease may be judged as devices for explaining and hopefully counteracting the inevitability of death, as well as socially rationalizing man's essential powerlessness and vulnerability. This, indeed, must be viewed as a fundamental and highly important (original) function of such taxonomies. In Western societies, on the other hand, a consequence of the growth of biomedicine and science in general has been the erosion of just such functions. Since theories of disease and explanations of their causes now bear no strictly *logical* relations to matters of death nor to social happenings *per se*, such theories cannot be used to socially rationalize death [5]. Important social concerns dealing with the latter questions have in principle been delegated to other institutions, and are rationalized by an appeal to non-medical ideas, ideas whose power, compellingness, and influence in social affairs could be seen as weakened as a result of their logical isolation from matters of disease.

There is an additional and interesting difference in the way non-literate taxonomies of disease are used as compared to the biomedical one. Among pre-literate people, disease (a category rendered by a practitioner by means of the taxonomy), illness (the experiences of the person said to be diseased), and maladaptions of an individual (defined by him or significant others as disruptions in function) are fused together. Each of these, in other words, implicates the other. On the other hand, in contemporary societies where the biomedical taxonomy of disease is found, these three conditions or entities are logically independent though they can and often do relate empirically. The development of modern medicine, in short, may be judged to have logically required the setting apart of these three facets of 'medical' experience which, in simpler societies, are interconnected. Thus, among non-literates, many such things as crime, socially disruptive behaviors, unusual sexual preferences, excessive reliance on alcohol or other drugs, and the like, are not judged as medical phenomena, but as habits or styles of living. In general, the issue of disease or illness is not ordinarily raised unless the individual is unable to look out for himself and his behavior conforms closely to a pattern which is culturally interpretable as non-motivated. In contrast to this, in the modern states, different forms of what are termed 'deviance' are often judged, first of all, as forms of maladaptation. Interestingly, they recently have come to be judged as forms of disease. The irony, of course, is that, in the nation states, disease has more and more become an abstract biological thing which, as stated above, on formal grounds is socially disconnected; whereas among non-literates, disease is defined purely on social behavioral grounds!

One can say then, that in settings where disease is defined behaviorally and is made a part of the social fabric, behaviors which we now take as forms of deviance or maladaptation infrequently are viewed as reflecting disease; on the other hand, in contemporary societies where disease has more and more been made a technical and non-behavioral thing, such behaviors often seem to raise the question of disease. It hardly needs emphasis that phenomena brought under the confines of a medical taxonomy entail predictable forms of social valuation, regardless of which type of social group this takes place in. Consequently, the preceding generalization points to the very different symbolic uses and functions which taxonomies of disease may have in social groups. On the one hand, the extension into social behavior of a taxonomy which more and more has come to explain chemical and physiological things may be seen as bringing into play earlier definitions of disease; on the other hand, since a new class of social behaviors is involved, one can say that the social domain of disease has expanded somewhat.

To summarize this and related points, one can say that, among preliterates, taxonomies of disease are linked of necessity to the matter of illness, death, finitude and the ultimate bases of social life, but not to those involving the everyday inconsistent and disruptive behaviors of co-members. In the nation states, on the other hand, a necessary link between death, disease, and illness does not exist, yet we nevertheless observe a tendency for this abstract and impersonal biomedical taxonomy to, at times, incorporate or explain a range of deviant actions heretofore thought of purely in social terms.

IX. THE CERTIFICATION OF MEDICAL TAXONOMIES

The question of what constitutes the identity and meaning of a disease form thus touches centrally on the whole problem of social control and organization. It also involves fundamental issues in biology, social systems, and the philosophy of science. In essence, social groups (or powerful sub-groupings in them) determine and certify what is to pass as a disease, what is its cause and/or when a cause of disease shall be said to be known or 'established.' For one, these groups articulate the relevant norms for disease marking, and the relevant definitions, ontologies, and boundaries of disease. Secondly, insofar as person centered maladaptations (variously constructed and perceived) constitute the raw data of disease, it is the successful control and/or elimination of this that is socially prescribed. This means, in essence, that, to a certain extent, properly understanding a disease is a matter of achieving proper control and/or elimination of the disease. But what stands for a proper cure or elimination is also a matter of definition and depends on the associated frame of reference and canons for evidence that hold sway in the group as well as their tolerance for disorder and disability, all of which relate directly to social values. Thus, for example, it could be said that we actually know more about multiple sclerosis (i.e., about the gross and microscopic anatomy, physiology, biochemistry, immunology, etc., of this disease) and about how and why changes associated with it are incapacitating than we do about, say, polio, smallpox or cholera (for example, how the respective pathogenic agents produce toxins, how the toxins affect man, and why these changes obtain). Still we say we know a great deal about polio and cholera, but not about multiple sclerosis. Perhaps this is the case because we are said to know the cause of polio, smallpox and cholera, but not of multiple sclerosis, where 'cause' is made to mean that we have achieved a satisfactory level of control.

The criteria for knowing and understanding disease, then, rest on criteria tied to a level of control that is deemed desirable on social grounds. For this reason, one can claim that the elegant experiments of Semmelweis and Graunt established satisfactorily *at the time* the causes of diseases such as 'child-bed fever,' 'plague' and 'pestilence,' even though, on the basis of hindsight, it is now 'known' that the identified agents, associated mechanisms of action and indeed disease types under consideration were poorly apprehended (see examples provided by Hempel [13] and Susser [22]). This is the case simply because at the time the experiments led to a form of social action which, in turn, brought about a desirable level of control and elimination of the 'diseases' in question. Later, such level of control was deemed unacceptable, so the efforts aimed at the finding of 'truer' causes were sanctioned.

In the last analysis, then, it is the social group that established the conditions, criteria, and grounds for judging what is disease and when the cause of disease may be said to be 'achieved.' Such judgments rest on social norms about what is noxious and also about what constitutes acceptable or tolerable control and/or elimination of the noxious. In this sense, then, social groups prescribe not only definitions of the ontology of disease but also of the standards of what shall pass as an acceptable understanding and control of that disease.

In Western societies, physicians, public health officials and researchers are the ones who pass judgment about disease. Their functions, although narrowly prescribed, are formally binding and they enjoy a monopoly in their domain or sphere of influence [18]. Judgments about cause and identity of disease occurrences are in principle, unequivocably established by physicians with whom we assume and share identical perspectives and modes of knowing. Among non-literates, shamans exercise a related function. Collectively, they can be said to articulate the veridical understandings about the causes and identities of disease. Their functions in their respective groups, however, are broad and diversified and moreover not necessarily binding in a formal sense. Furthermore, given the relatively low level of control over disease and the informal nature of the system of disease accounting that prevails, one might anticipate alternative perspectives which in turn produce equivocal and conflicting understandings about disease and their causes. The result of this is that on the whole there may exist a smaller degree of consensus about disease-related happenings [7].

X. TAXONOMIES OF DISEASE VIEWED FROM THE STANDPOINT OF SYSTEMS THEORY

Disease constitutes one of the elemental problems which any group must deal with and this is especially the case among non-literates. This may be the reason that in these groups, taxonomies of disease rest on religious premises. Anthropologists interested in the question of social adaptation have thus been forced to study the functions which the idea of sanctity has in social groups [20]. An important and relevant link is thus established between these four sets of factors: disease, taxonomies of disease, sanctity, and social adaptation. Taxonomies of disease also involve what others have termed the 'cognized' models of peoples; that is, the model which they have of the world and in terms of which they manage their affairs. Cognized models are to be distinguished from the 'operational' models of the analyst [2]. The latter, in applying his model, is interested in understanding how groups function and maintain themselves in their distinctive setting. The actions of people, performed in the light of their cognized model, have determinate effects on processes and happenings which the analyst describes in line with general systems theory [14]. The attempt is made to evaluate the functional consequences of group actions in terms of processes which the analyst identifies in his systems interpretation. The following question can be posed: How might a people's medical activities be approached from a systems frame of reference?

One can begin with the 'system' which is descriptive of the person. In general systems theory, the person is described as a composite of open, connected, and hierarchically ordered biologic (e.g., chemical physiologic) and psychologic systems which, in turn, connect with any number of social systems. These 'systems' implicate processes and happenings which are studied in the social and biological sciences. In order to interpret the person's actions and behaviors, his view of himself and of his situation needs to be represented. This raises the question of personhood and that of the actor's cognized models of his world, all of which bear a relation to the group's taxonomy of disease. The set of 'systems' which the analyst uses to describe the person thus includes one which articulates how he defines himself in the context of problems tied to a disease. As I have indicated, in this subsystem (i.e., the person's cognized model) the units implicated are not exclusively 'inside' the person nor necessarily of a 'lower order' as they might be in the observer's model of the same person. For example, persons can describe themselves in terms of souls which are located in sacred mountains, spirits which are drawn from animals or streams, as well as 'parts' which are located inside of his body. Which of these components are of a higher or lower order and which include the other is problematic, as is the matter of how the person stands in relation to other aspects of the environment to which he sees himself as necessarily related.

In short, the way in which the person describes himself bears little resemblance to how the analyst describes him. It is to be emphasized that it is the *analyst's* hierarchical model of the person and of his life situation which is used to understand and make sense of the person's actions. These actions may be judged as 'regulated by' developments in system happenings which are actually unbeknown to the person but which are nonetheless affected by him through actions undertaken in line with his own cognized model. One assumes that actions performed in terms of the rationale of cognized models can be interpreted and evaluated in the light of the requirements of systems theorists: this, in essence, means that systematic relations can be established between cognized and operational models.

Viewed from the standpoint of the observer's model, how the whole 'system,' which is the person, functions (i.e., adapts, gets diseased, gets treated, etc.), is an outcome of its connection with and control by a higher system which incorporates him. One could term this a 'medical' system. Together with other systems at the 'same level' which also implicate the person and contribute to his welfare (e.g., agricultural, familial, legal, etc.), the medical system constitutes but a subsystem in a higher 'system,' perhaps that of the cultural group of the people which must be seen in relation to other cultures of the region. These, in turn, are contained in a supra or higher order system which comprises the population as a whole, which in turn is located in the ecosystem, etc. . . . This characterization, to re-emphasize, constitutes the researcher's *operational model* which he uses in order to study questions involving the adaptation of the group and the problems and influences of human disease.

An analyst who wishes to study the function or value of medical taxonomies using the frame of reference of systems theory must naturally first adopt an independent measure of disease and of what generates disease. Given this, he might address, as an example, the functions or consequences of preternatural suppositions. An attempt would be made to examine whether such suppositions generate medically efficacious (or deleterious) actions. This would be done because preternatural suppositions often ground (i.e., more explicitly, sanctify) the medical activities of a people. Such suppositions are implicated in the taxonomy of disease which is an integral part of the cognized model of the individuals. The researcher interested in studying the adaptive value of the group's taxonomy of disease might be interested in the health status of the person or in the efficiency in the way the group functions as a whole, to use two examples of independent criteria about disease. In the former instance, sociobiologic medical data may be needed; in the latter, data about social relations and social organization. In either case, what the researcher would hope to show is that group rationalized behaviors which reflect the cognized models of the people, somehow contribute to the adaptation of that group as he views this in the light of his operational model.

A taxonomy of disease, then, is a special type of multileveled 'cognized' model. This model describes the actor and his world, it expresses his rationale for being in the world, and it provides him with a means of explaining and rectifying problems posed by disease, however he defines these. Just as ethical and religious directives inform medical taxonomies or cognized medical models – perhaps even rationalize and justify them – so also related matters of sanctity and the effects of religious propositions have been judged to explain the function and adaptation of social groups[20]. Elsewhere I have used this approach to study the function of medical care systems [8]; also see [6].

XI. TAXONOMIES OF DISEASE AND THEIR RELEVANCE TO PSYCHIATRY

Characteristics of taxonomies of disease have been reviewed. Implicit in the discussion has been the notions of comparison and differences. Taxonomies of disease may also be held to grow and change. A review of medical orientations of non-literate, early Western and modern social groups suggests that the emphasis placed on the various regions of the taxonomy of disease changes. Symbols which refer to supernatural influences diminish in frequency and symbols which refer to the region of the worldly environment and especially of the person become the dominant vehicle for medical explanations. To one interested in the study of human behavior and of disease, this shift in emphasis is important and has obvious implications: people will more and more come to rationalize and deal with disease through the use of symbols which articulate the newly posited features of these regions. New social realities come to be established and then represented in the taxonomies. One could claim that the sources and foundations of the changes experienced by an individual who is diseased do not change. What does is his reading or interpretation of those changes. With this, come obvious changes in behavior.

A relatively recent development in the history of medical taxonomies, a development inextricably linked to the whole problem of modernization and the growth of modern science, has been that taxonomies have made reference to phenomena which are more and more separated from the social behaviors of persons and indeed from their social relations. The process whereby indicators of disease come to be disconnected from social behavior has been associated with the separation of the idea of illness from that of disease, and moreover, both of these from maladaption. The notion of an individual having a disease in the absence of being ill, a notion made possible by abstracted 'invisible' features of persons, has allowed members of groups to use their taxonomy to link together episodes of illness which are separated in time. We thus are provided with chronic and subclinical diseases. In non-literate groups as implied earlier, medical taxonomies are used to explain occurrences of diseases (i.e., episodes of illness) which are seen as unitary and continuous eventuations. This literally means that disease exists only while the individual is ill; it also means that the resolution of that disease as illness *terminates* the question of the individual's impairment in competence and capacity. However, the development of symbolizations which refer to entities abstracted out of behavior and, as it were, hidden in the mind and/or of the person, allows for the question of competence and capacity (i.e., disease) to be raised in the absence of illness. This means that it has now become possible to use the idea of disease to make enduring characterizations of persons which derive their authority from a scientifically based medical taxonomy.

I am suggesting, in short, that dualism, and associated themes which now, as it were, serve to partition the region of the person in our taxonomy of disease, are relatively recent inventions in the history of man as a species. Dualism has allowed us a new way of looking at ourselves and the world. In fruitfully channeling observation and research it has been instrumental in the growth of biomedicine. At the same time, dualism and biomedicine have produced a number of paradoxes and dilemmas; and medicine in general and psychiatry in particular are ensnared by them.

At this point in the evolution of modern medicine, psychiatry occupies an anomalous position and I would like to suggest two interrelated factors which contribute to this state of affairs. First of all, in contrast with other contemporary medical disciplines whose disease indicators have become largely disconnected from social behaviors – to be identified and measured in chemistries, tissue and X-ray shadows – psychiatry's indicators of disease are still rooted in behavior. Because this medium is still critical in the definition and recognition of psychiatric disease, all influences which affect it need to be taken into consideration in diagnosis. In a word, psychiatry seems not yet able to share in the luxury of neatly abstracting out its indicators of disease from the tangled, enigmatic and ambiguous webwork of social behavior. This point can be driven home very easily. One can train an individual to behave as a patient with a myocardial infarction or as one affected by a brain tumor. However, our actor would be found out easily enough, and that is because we have laboratories, X-ray machines and established principles about how the nervous system functions which cannot be easily learned by our actor. On the other hand, we could also train an individual to behave as one showing schizophrenia or depression. In this case, full training would seem to easily have him appear schizophrenic. Moreover, a reasonable bet would have it that he might *not* be found out. Some would even claim that a complete enough indoctrination would be enough to make him schizophrenic or depressed. The point here is simply that the raw stuff of diagnosis in psychiatry is social behavioral changes, that this medium constitutes *at present* the principle criteria of disease, and that this distinguishes psychiatry from other contemporary medical disciplines. There is indeed an irony in this: in endorsing the whole of the person and in particular his behavior – something that seems integral to the idea of disease when one adopts a comparative and evolutionary perspective (see [5, 6]) – psychiatry can be criticized as being unscientific and old fashioned.

The second reason psychiatry occupies an anomalous position in biomedicine stems from the obvious fact that its locus of concern happens to be the brain and its mode of operation, and that one of this organ's functions is, in fact, to regulate social behavior. This is a basic generalization that psychiatry itself has helped establish. The more we learn about the brain, how it functions, and about its growth and development, the more we learn about how many different kinds of influences are required for its development; among these, cultural rules and guidelines which are integral to socialization and development of the early organism are all-important.

This paradox, then, is at the core of psychiatry and needs to be appreciated. On the one hand, the discipline has thus far been unable to diagnose disease by procedures and methods that are independent of behavior. Its basic medical tasks of diagnosis and treatment require, at present, that careful attention be given to social behavioral changes and this necessarily means cultural influences. This separates psychiatry from other disciplines. On the other hand, the focus of concern of psychiatry is an organ, the function of which is in fact to regulate social behavior. Disturbances of this organ will inevitably be expressed in changed social behaviors. Ironically, to fully understand how this organ functions chemically and psychologically one needs to appreciate and take into consideration the role of social and cultural influences. Brain function cannot develop in a beaker bathed in a mixture of nutrients but rather in behaving, adapting persons who are influenced by specific social happenings and cultural symbols. Moreover, the brain itself is a product of evolutionary processes which themselves have been importantly affected in a feedback relationship by both cultural and neurobiological factors – all of which are realized in social behavior [24, 17].

There is an additional aspect of the way the psychiatric taxonomy of disease functions and is used which bears examination. To begin with, recall that ordinarily two related conditions need to be met in order for 'disease' to be properly said to characterize a person: he or she manifests impairment or disability, and at the same time reports or shows a number of interferences in basic functioning. (In our taxonomy, of course, disease can apply when we have reason to anticipate that these interferences will follow and this creates special problems.) Examples would include interferences with such things as walking, talking, digesting, defecating, urinating, sensing, moving, etc. . . . To these must be added that of perceiving, thinking and rule following. The latter three one can think of together as a 'social triad' which makes the content of behavior meaningful and predictable to co-members. Examinations of the medical orientations and practices of diverse people, whether primitive vs. modern, or Western vs. Non-Western, discloses that a subset of these interferences usually tend to be associated with the claim of disease. In the case of non-literates, all of these interferences, *except the triad* which includes rule following, are interpreted in an emblematic way; this is to say, their presence in or their report by a person is taken to *directly* validate the question of disease. The social triad which includes rule following, however, is not interpreted in this direct and emblematic fashion; instead, the motives or posited reasons behind the break in the social triad become the focus of analysis. It is the social meaning or sense which is given to this latter subset of interferences which establishes whether the presence of disease (as opposed to criminality, witchcraft, malevolence, drunkenness, etc. . . .) is the relevant category for explaining the interferences.

As mentioned earlier, in modern societies where the biomedical taxonomy certifies medical activities, one notes a converse state of affairs. Here, recently the social triad has come to be assessed in a direct and emblematic way, with the question of motives or social meaning of behavior blurred or muted. On the other hand, in our taxonomy, and in contrast to non-literates, the relevance of the *meaning of* or *motives behind* the remaining human interferences (i.e., interferences with talking, defecating, sensing, etc. . . .) are now considered relevant. This, in fact, is why we have entities such as hysteria which in contemporary biomedicine is viewed by many physicians as a spurious disease. Hysteria is a disease which raises the question of motives or alternative meanings behind a host of interferences which among non-literate groups are directly taken to indicate disease (i.e., are judged emblematically). Among non-literates, in other words, 'hysterical' diseases do not and could not exist, and the reason for this is that the question of what is behind the 'non-social' interferences does not arise as a possibility. The idea of a conversion symptom cannot arise since taxonomies cannot get behind symptoms and assign physiological truth values to them. In a sense all symptoms are converted from their cause; there is no hierarchy of authenticity such as seems inherent in biomedicine. There is thus an interesting contrast in the way basic interferences in functioning are assessed in these two types of societies.

The identification which psychiatry has and continues to have with 'hysterical disease' may thus be viewed as reflecting and contributing to its dilemma in contemporary medicine. This dilemma stems from the fact that in contrast to other medical disciplines, its conception of disease is rooted in a mentalistic interpretation of human functionings. This allows it to claim that hysteria is a legitimate disease, a claim that can only be sustained in light of the very varied kinds of highly abstract symbolizations of personhood which prevail in the biomedical taxonomy of disease. The irony of this is, again, that the dilemma follows logically from psychiatry's traditionally medical concern of viewing disease and persons holistically. In other words, psychiatry is the medical discipline which directly addresses social behavior and adaptation for definition of its domain of interest – something medical taxonomies have until recently *always* done.

A final observation may be in order. From the standpoint of contemporary critics of psychiatry, hysteria is a paradigm psychiatric concern [21]. Their interpretation, however, draws mainly on the question of personal responsibility and the dependency made possible by the sick role, both of which follow from their particular rendition of psychoanalytic ideas. In the formulation presented here, hysteria remains a paradigm case. However, its position and importance are viewed in the broader context of the evolution of taxonomies of disease and social system generally. Nevertheless, both interpretations of the problem of hysteria underscore the anomalous position of psychiatry in contemporary medicine.

University of Pittsburgh,
Pittsburgh, Pennsylvania

REFERENCES

1. Berlin, B. and Kay, P.: 1969, *Basic Color Terms*, University of California Press, Berkeley and Los Angeles.

2. Caws, P.: 1974, 'Operational, Representational, and Explanatory Models', *American Anthropologist* **73**, 59–76.
3. Cole, M. and Scribner, S.: 1974, *Culture and Thought: A Psychological Introduction*, John Wiley & Sons, New York.
4. Fabrega, H., Jr. and Silver, D.: 1973, *Illness and Shamanistic Curing in Zinacantan: An Ethnomedical Analysis*, Stanford University Press, Stanford, California.
5. Fabrega, H. Jr.: 1974, *Disease and Social Behavior: An Interdisciplinary Perspective*, MIT Press, Cambridge, Massachusetts.
6. Fabrega, H.: 1975, 'The Need for an Ethnomedical Science', *Science* **189**, 969–975.
7. Fabrega, H.: 'The Biological Significance of Taxonomies of Disease', to appear in *Journal of Theoretical Biology*.
8. Fabrega, H.: 'The Function of Medical Care Systems: A Logical Analysis', *Perspectives in Biology and Medicine* **20** (autumn 1976), 108–119.
9. Frake, C. O.: 1961, 'The Diagnosis of Disease Among the Subanun of Mindanao', *American Anthropologist* **63**, 113–132.
10. Geertz, C.: 1973, *The Interpretation of Cultures*, Basic Books, New York.
11. Globus, G.: 1973, 'Unexpected Symmetries in the "World Knot"', *Science* **180**, 1129–1136.
12. Goodenough, W. H.: 1970, *Description and Comparison in Cultural Anthropology*, Aldine, Chicago.
13. Hempel, C. G.: 1966, *Philosophy of Natural Science*, Prentice-Hall, Englewood Cliffs, New Jersey.
14. Laszlo, C. A., Levine, M. D., and Milsun, J. H.: 1974, 'A General Systems Framework for Social Systems', *Behavioral Science* **19**, 79–92.
15. Lounsbury, F. G.: 1969, 'Language and Culture', in S. Hook (ed.) *Language and Philosophy*, New York University Press, New York.
16. Mooney, J. and Olbrechts, F. S.: 1932, 'The Swimmer Manuscript: Cherokee Sacred Formulas and Medicinal Prescriptions', *Bureau of American Ethnology, Bulletin 99*.
17. Napier, J.: 1973, *The Roots of Mankind*, Harper, New York.
18. Parsons, T.: 1951, *The Social System*, Free Press, Glencoe, Illinois.
19. Quine, W. V. and Ullian, J. S.: 1970, *The Web of Belief*, Random House, New York.
20. Rappaport, R. A.: 1971, 'Ritual, Sanctity, and Cybernetics', *American Anthropogist* **73**, 59–76.
21. Szasz, T.: 1976, 'The Concept of Mental Illness: Explanation or Justification?', in this volume, pp. 235–250.
22. Susser, M.: 1973, *Causal Thinking in the Health Sciences: Concepts and Strategies of Epidemiology*, Oxford University Press, Oxford.
23. Turner, V. W.: 1963, *Lunda Medicine and the Treatment of Disease*, Government Printer, Lusaka, Northern Rhodesia.
24. Washburn, S. L. and Harding, R. S.: 1970, *Evolution of Primate Behavior in the Neurosciences, Second Study Program*, edited by F. O. Schmitt, Rockefeller Press, New York.

RUTH MACKLIN

HEALTH AND DISEASE: THE HOLISTIC APPROACH

I

The attempt to view disease in a generic sense – as an attribute of living forms, with special focus on its adaptational significance – is an approach that deserves acclaim and further study. It is somewhat difficult, however, on first hearing Dr. Fabrega's position on the importance of viewing disease as a symbolic category, to see the connections between this point of view and current debates about how to conceptualize and categorize mental health and mental illness. It is perhaps easier to imagine how the development of what he calls 'ethnomedical science' [5] might contribute to the ongoing debate about whether there are any universal characteristics of illness – mental or otherwise – or whether the nature and varieties of human diseases are relative to the cultures in which they exist. But a close look at the details of Dr. Fabrega's position reveals several links between his proposal to view disease in a generic sense and the development of an adequate theory of mental health.

First, his proposal to view disease as a symbolic category – as a feature of living forms – places a needed emphasis on the adaptational significance of disease states and their various manifestations. Second, this viewpoint supports a holistic conception of health and illness as states or attributes of persons, rather than of bodies or minds, thus undercutting the misleading dualistic conception of human beings that has prevailed both in Western philosophy and in medicine. Third (perhaps as a consequence of considerations noted in the first two points), the role of psychological factors as partial determinants of health and illness emerges from accounts such as this in a way that strengthens a monistic or unitary conception of persons and in so doing, renders unnecessary or misconceived a variety of conceptual attacks on the coherence or meaningfulness of the notions of mental health and illness.

So while the direct links between Dr. Fabrega's account and the topic of mental health may not be immediately apparent, I think the connections are significant ones, worth making explicit. Since I am in essential agreement with Dr. Fabrega concerning the overall importance and heuristic value of a holistic approach, I want to begin by affirming this basic accord. I shall have

H. T. Engelhardt, Jr. and S. F. Spicker (eds.), Mental Health: Philosophical Perspectives, 107–117.

some critical things to say about the details of his position, but these criticisms are not leveled at the underlying presuppositions or the structure of his conceptual framework. Instead, my critical remarks aim to show that Dr. Fabrega's view might be sharpened, refined, or modified so as to fit better into the holistic framework he has chosen to promote. My chief criticisms are (1) that he neglects or rejects – in what I argue is an unnecessary or unwarranted dismissal – the notion of causality in the holistic framework; (2) that he overemphasizes the practical importance of the logical and semantical features of the *concept* of disease, at the expense of people's *beliefs* about causation of disease and methods of preventing and combating illness; and (3) that the approach he takes to the issues in (1) and (2) impedes his own acknowledged attempt to reunite mind and body into the psychobiological unity perceived by other cultures but torn asunder by Western biomedicine in the Cartesian tradition. Since I support Dr. Fabrega's theoretical approach, I hope to assist in its development by addressing my remarks to those features of his account that might be modified so as to contribute to his own acknowledged aims.

II

Before launching into these three arguments, however, I want first to say a few words about holistic approaches to disease generally and to psychological factors, in particular. This will serve not only to explicate the holistic framework more fully, but also to lay the groundwork for the modifications I later urge for Dr. Fabrega's account. In a recent article, he asserts his own conception of what this conceptual framework entails, as follows:

> . . . one needs to begin to look at disease not only as reflecting psychological conflicts or narrowly defined physiological changes, which have importance in the light of an individual's immediate past (and with the exclusive aim of uncovering causes for treatment), but also as reflecting other more basic factors about man in relation to social systems. This broad and essentially holistic approach to disease and to the question of human adaptation is implicit in this analysis ([6], p. 1501).

A holistic conception, in the most general sense, is one that structures the theory and practice of the domain to which it applies, and is perhaps best thought of as a metaphysical schema. It is metaphysical in the sense used by contemporary philosophers, that is, as providing a conceptual framework for structuring and ordering data, according ontological status to objects, states, or events, and addressing problems of individuation and identity. Issues in scientific reductionism are metaphysical in this sense: can psychological states and processes be 'reduced' to neurophysiological or biochemical states and processes (are there two sets of states and processes or really only one)? Can macro-events at any level (e.g., sociological, behavioral, bodily) be reduced to micro-events (brain processes)? We need not address these questions here, but it is important to bear in mind the sorts of questions raised by conceptual frameworks – holistic or otherwise.

One of the best known and most compelling proponents of the overall value of a holistic approach to health and disease is René Dubos. In his words:

> Because of the peculiarities of the human condition, the problems posed by health and disease often involve a variety of facts which appear far removed from those studied by natural scientists. Indeed, medicine would soon lose much of its relevance to the welfare of man if it were to limit itself to problems that can be analyzed by the orthodox techniques of physicochemical sciences. . . . It is almost certain, in fact, that medicine will eventually flounder in a sea of irrelevancy unless it learns more of the relations of the body machine to the total environment, as well as to the past and the aspirations of human beings. . . . What is needed is nothing less than a new methodology to acquire objective knowledge concerning the highest manifestation of life – the humanness of man ([3], pp. xix–xx).

A holistic theory, then, is one that takes the whole person, viewed in connection with his life-space, as the proper unit of analysis according to which health and disease are to be understood. It is fully compatible with the existing medical sciences, since a holistic framework views data and theories from medical specialties as one relevant concern, among others, for the understanding and explanation of human health and illness.

This approach is one that necessarily views man as a psychobiological unit, not as a composite of mind and body. This rather simple but highly significant conceptual shift enables us to make better sense out of psychosomatic illnesses [1] and the all-important phenomena of psychological factors in healing (e.g., the placebo effect, religious beliefs, faith in the medical practitioner or healer) [7]. Moreover, if the unified conceptual framework no longer allows us to speak of diseases of the mind and diseases of the body, then this should undercut – at least in part – the arguments given by Thomas Szasz and others against what they take to be a 'logical' error in treating so-called 'problems in living' as mental illnesses [10, 13]. Sometimes people have physical ailments, which give rise to or are accompanied by psychic distress; sometimes they experience anxiety, tension, or depression – so-called mental states, but states that nonetheless have invariable physical and behavioral accompaniments of one sort or another. On a holistic conception, the terms 'mental' and 'physical' are not to be taken literally as denoting essentially different substances, but they can still be retained as convenient labels (without pain of contradiction) to refer to matters of emphasis, or most prominent presenting symptoms, or roles in a complex causal nexus.

There is increasing agreement, even among theorists espousing diverse psychological viewpoints, that so-called mental health and illness – whatever their proper conceptual analysis – are directly affected by social and environmental factors. These factors include a wide variety of circumstances pertaining to family, occupation, life-style, stress, life changes, and numerous other conditions comprising every individual's social milieu. Within the fields of psychiatry and psychology, there have been several spokesmen for a holistic conception of the human personality – often known as organismic theory ([8], pp. 298–337). Such theories emphasize the 'unity, integration, consistency, and coherence of the normal personality' ([8], p. 300). Organismic approaches to personality are most often instances of the type known as self-actualization or self-realization theories, and as a result, they usually express a positive concept of mental health instead of the viewpoint that conceives of health as the absence of disease, or places health and illness at opposite ends of a continuum [9, 11]. The organismic or holistic conception maintains, in sum, that 'it is impossible to understand the whole by directly studying isolated parts and segments because the whole functions according to laws that cannot be found in the parts. . . . Organization is the natural state of the organism; disorganization is pathological and is usually brought about by the impact of an oppressive or threatening environment, or, to a lesser degree, by intraorganic anomalies' ([8], p. 330). Theories that are not holistic in this sense will most likely be inadequate to the task of accounting for the complex etiology of disease generally, as well as to the more limited job of explaining particular instances of falling ill or of malfunctioning behavior.

Having laid this basic groundwork by noting some features of the holistic approach, let us turn now to the first critical contention about Dr. Fabrega's account.

III

In his explication of disease as a symbolic category, Dr. Fabrega has chosen to neglect or reject the relevance of the notion of causality for the holistic approach he adopts. I shall argue that this dismissal of causality is either unnecessary or unwarranted; that is, acknowledging the role of causal factors is not incompatible with viewing disease as a symbolic category and, moreover, it can contribute further to our understanding of health and disease as states or conditions of entire organisms.

As evidence for my claim that Dr. Fabrega disavows a primary interest in the truth of causal hypotheses about disease, I take the following statements:

I have tried to approach disease largely in terms of the cultural beliefs and orientations which guide people to various forms of 'medical' action. This emphasis on the cultural medical perspectives of different people is what I mean by the title of my paper, "disease as a symbolic category." ... [A]ll types of beliefs or theories of disease are [then] relevant, regardless of how developed or simple one may judge the group to be. This is so because the phenomenon, disease or illness, is viewed in terms of the group's system of signification ([4], p. 80).

Later he claims:

I must reiterate that the accuracy or correctness of the way the group divides up the region of the person in explaining disease is not important in this phase of an analysis of a group's system of medicine. What is important is simply how and also the degree to which the region is specified and articulated as a phenomenal structure ([4], p. 86).

And further:

... one may disregard the notion of the validity of the explanation – the so-called 'scientific' truthfulness of either of its premises and/or key concepts. We can focus instead on its properties as a symbolic system, on the degree of specificity and differentiation of the symbols, and on the degree of intricateness in the way they are used to produce appropriate explanations of disease ([4], p. 88).

These remarks make it appear that Dr. Fabrega's concern is not with the scientific accuracy or objectively determined truth of different cultures' symbolic representations of disease, but rather, with how such perspectives function within a society from an anthropological or social scientific point of view. But making *social scientific observations* of this sort is not incompatible with the activity of *making scientific assessments* of the accuracy or validity of a culture's theory of disease or explanatory schema. Indeed, both tasks together may well serve to enhance our own scientific understanding of human health and disease. Dr. Fabrega appears to conflate these two different activities, or else fails to recognize that simultaneously undertaking both is a possible and fruitful way of deepening our understanding of these matters.

Why, then, should one disregard the scientific truthfulness of any proposed explanation of disease? Why should one focus *instead* on its properties as a symbolic system? If one seeks a value-neutral description of a culture's beliefs about disease, one can still judge the truth or falsity of those beliefs by reference to known scientific principles and canons of inquiry. At the same time, it would be a mistake to dismiss the beliefs of a culture or of individuals as irrelevant to the processes of health and disease. The virtue of a holistic approach is that it requires us to look for causal factors and conditions in places that the more fragmented methodology of Western scientific medicine has until recently overlooked. As Jerome Frank demonstrates at length in his book *Persuasion and Healing* [7], psychological factors operate as important causes or conditions in the way diseases take their course or healing occurs. Frank draws on data from non-literate cultures that employ shamans, on the practice of religious healing generally, as well as demonstrating the role of beliefs in a variety of psychotherapeutic contexts. These he links with the well-known phenomenon of the placebo effect in healing, and argues persuasively for the significance of the common features among all of these phenomena, including also Communist thought reform.

There appears, then, to be no good reason to ignore the scientific truth or accuracy of a proffered explanation, since our own understanding of causal factors or conditions that promote or retard healing may well be deepened by observations of successful healing among non-literate cultures. The knowledge that some of their specific beliefs and proffered explanations about disease are materially false may serve to make us aware of factors contributing to health and disease that we otherwise might have overlooked.

Dr. Fabrega himself seems to envision these sorts of goals as outcomes of the inquiry into taxonomies of disease. He notes that

> taxonomies of disease . . . reveal a group's symbolizations about elemental matters of both intracultural and universal significance. When they are studied in conjunction with suitable (independent) languages of disease that specify properties and components of (a) disease(s), they may provide a means of addressing matters of central concern to biology as well as social science ([4], pp. 89–90).

This demonstrates important links with the concerns of Western medical science, but underlines the need for an evaluation of the content and correctness of the culture's belief system about health and disease. A greater emphasis on the role of causal factors is more likely to lead to results of 'universal significance' than is the dismissal of their importance when studying the symbolic function of disease. Since the symbolic approach and the causal approach are not incompatible and can proceed simultaneously, their combined use in cross-cultural studies can only enrich the insights afforded by a holistic conceptual framework.

IV

Turning now to my second critical contention, I shall argue that Dr. Fabrega's account overestimates the practical importance of the logical and semantical features of the *concept* of disease, at the expense of considering people's *beliefs* about causation of disease and how these beliefs enter into methods of preventing and combating illness. Let us begin with a look at an extended passage in which he sets out his position:

> . . . any episode of illness explained by this taxonomy does not directly draw on or feed into social and psychologic matters. One can say that in this particular taxonomy – the biomedical taxonomy – diseases reflect impersonal and technically malfunctioning sorts of things and are literally socially *nonsensical*. Furthermore, . . . in biomedicine once a diagnosis is reached, then *regardless* of the differing courses that a disease may pursue, it still tends to have the *same fixed* significance which remains uncoupled logically from social affairs. . . . To put the matter succinctly, little of what cancer, tuberculosis, or myocardial infarction portend to persons or to the group can be related to social processes and relations by an appeal to the biomedical taxonomy qua system of meaning ([4], p. 92).

In this passage and elsewhere in his paper, Dr. Fabrega makes a series of observations about the logical features and semantical implications of the biomedical taxonomy, noting what he claims to be their logical independence from social, psychological, and cultural factors. But the significance of this observation is not made convincing, especially when we realize that even if occurrences of disease labelled according to the biomedical taxonomy may not *logically* portend social and psychologic matters, still they may *causally* portend them, which is more to the point. When medical and lay persons alike recognize and take account of the role these latter factors play in the propensity to fall ill or to heal rapidly, the semantic features of the taxonomy would appear to be of little significance.

It is, then, not clear to me why Dr. Fabrega attaches so much importance to the semantical features of the biomedical taxonomy or to disease labels and their logical implications. It may indeed be true that the logical features of the prevalent germ theory of disease have led to distorted conceptions of disease generally and, in particular, to the acceptance of an oversimplified causal model of human disease. However, the problem does not seem to lie in the *semantical* features of the taxonomy, but rather, in the empirical and methodological errors arising out of the use of the overly simplistic germ theoretic model. While I think I am in basic agreement with the point of view that motivates Dr. Fabrega's criticism of the Western biomedical taxonomy, I believe he is mistaken in imputing to semantical features of the taxonomy what should instead be attributed to professional and lay beliefs about the role of psychological and social factors in cause and cure of human disease.

In support of my contention, let me focus briefly on the somewhat recent awareness on the part of both medical practitioners and lay persons of the role of stress as a causal condition in both mental and physical illness. It is not only the work of Dr. Hans Selye [12] on the complex and diverse effects of psychological stress and its contributory role in diseases of various sorts that is instructive here. There is also the recently awakened interest in stress reduction by such techniques as transcendental meditation or yogic meditation – techniques that would most likely have been dismissed as recently as a decade ago by practitioners of Western scientific medicine as excursions into Eastern mystical nonsense. Interest in the physiology of meditation and related techniques of stress and tension reduction (e.g., biofeedback) has not only been manifest in reputable publications like *Science* [14] and *Scientific American* [15], but also been evident in the popular press with books such as *The Relaxation Response* [2]. So it is not only physicians – many of whom now seek causes of elevated serum cholesterol in the life style of persons rather than in amounts of butter and eggs ingested – who have expanded their belief system about causes of disease and modes of healing. It is also the public at large, made aware of the relationship between stress and disease, who flock to courses in TM or do yogic exercises, who seek desensitization training of various sorts and increasingly attribute many of their physical ailments to social and psychological factors. Even if these factors remain *logically* or *semantically* divorced from specific disease concepts, as Dr. Fabrega claims, this logical independence seems of little *practical* import for the beliefs and actions of people regarding cause, cure, and prevention of illness.

V

The first two criticisms that I have leveled against Dr. Fabrega's account seem more to be questions of preferred emphasis than of substantive disagreement. The third point of discussion is not so much an additional criticism as it is an observation that arises out of my first two points. Here I want to argue that the approach Dr. Fabrega takes to the issues discussed in my earlier criticisms impedes his own acknowledged attempt to consider persons as mind-body wholes rather than as Cartesian dualistic entities. If we focus more on the causal (rather than the symbolic) factors relating to disease categories, and if we emphasize belief-systems about social and psychologic factors in disease (rather than the logical implications of disease concepts), then we are in a better position to give the holistic conceptual framework its proper due as an *explanatory* model for human health and disease. Not only are we better able to assess the importance of psychological factors in healing – thus lending further support to a view of man as a psychobiological unit; but we may also reexamine the by-now overworked issue of the similarity and difference between so-called mental illness and so-called bodily illness. The more we can provide a cogent, scientifically sound conception of persons as psychobiological units inhabiting a life-space that is relevant to their health and illness, the more blurred the lines become between physical and mental disease.

Dr. Fabrega describes his approach to mind-body issues as follows:

> . . . the term 'body', as ordinarily used, is tied to our own Western philosophic premises and draws part of its meaning from the contrastive category 'non-body,' ordinarily phrased as 'mind.'. . . To non-literates and non-Western people (and probably to 'man' seen in an evolutionary time frame), dualism and the idea of an impersonal and mechanically functioning 'body' in the context of disease has little meaning . . . . [I] t seems prudent for now not to rely on the Western dualistic approach which tends towards reductionism ([4], pp. 81–82).

But now, we need to ask, what are the shortcomings of the dualistic conception of persons that stand to be remedied by a holistic approach? Surely it cannot simply be that non-literates and non-Western people have not formulated a dualistic conception, so Western medicine also ought not focus on the mechanically functioning body in the context of disease. It must be that our understanding of health and disease will be increased or that our ability to explain and predict occurrences of illness will be improved by abandoning what is perhaps best described as an artificial dichotomy between mind and body. In the same light, we need to ask: what is wrong with reductionism as a methodological or theoretical precept? It is unlikely that we shall find anything inherently mistaken or misconceived in reductionist approaches in science generally. The acceptance of tenets of scientific reductionism has flowed from the search for 'deep' explanations of natural phenomena, as well as from an adherence to the principle of parsimony or use of Occam's Razor in scientific theorizing. If this is so, reductionism itself should not be viewed as the villain in biomedical or behavioral contexts. Rather, it is the inability of the reductionist approach to afford complete and accurate explanations of the phenomena of health and disease that makes its use inappropriate in medicine – as opposed, say, to a field such as physics. There seems, then, to be a central epistemological consideration – one that not only throws light on non-Western conceptions of diseases and their alleged causes and cures, but also brings together insights from the fields of psychiatry and psychosomatic medicine, inquiries into the relationship between stress and disease, and studies of people's life changes and the frequency and severity with which they become ill [12, 16].

If an organism is subjected to undue stress from the physical, psychological, or social environment, the failure to adapt may manifest in physically measurable conditions such as hypertension or development of lesions like peptic ulcers. Or it may manifest in psychological or emotional states such as anxiety or nervous tension. Or it may manifest in states or conditions on the borderland between the mental and physical such as headaches or feelings of exhaustion. If similar sets of antecedent conditions and types of events are found repeatedly to give rise to some sort of malfunctioning of the organism – whether we identify the malfunctioning as physical or mental or some of each – this finding would seem to lend strong support to an organismic conception of persons. Maladaptive responses of the organism may take the form of organic, emotional, or behavioral states or processes; there seems little reason to select some subset of these responses as that which properly constitutes disease. Not only would we be forced to make artificial distinctions between the mental and the physical if we had to classify every maladaptive response of an organism into failures of mind and failures of body; but we would also overlook much that is causally relevant in the processes of disease and healing.

If it is difficult or impossible to sort out bodily from non-bodily maladaptive responses of organisms, then it appears equally difficult – if not pointless – to try to include only bodily malfunctions as instances of disease, relegating others to the realm of 'problems in living' or some other vague category. If persons are correctly to be viewed as psychobiological units, then one can no more mark off the class of bodily malfunctionings and call them diseases than one can find uniquely 'mental' forms of maladaptive responses. The approach shared by Dr. Fabrega and others constitutes a gain in conceptual clarity as well as in theoretical accuracy and explanatory power in the domain of human health and disease.

Case Western Reserve University,
Cleveland, Ohio

BIBLIOGRAPHY

1. Alexander, F.: 1950, *Psychosomatic Medicine*, W. W. Norton & Company, New York.
2. Benson, H.: 1975, *The Relaxation Response*, Morrow, New York.
3. Dubos, R.: 1965, *Man Adapting*, Yale University Press, New Haven.
4. Fabrega, H., Jr.: 'Disease Viewed as a Symbolic Category', in this volume, pp. 79–106.

5. Fabrega, H., Jr.: 1975, 'The Need for an Ethnomedical Science', *Science*, **189**, 969–975.
6. Fabrega, H., Jr.: 1975, 'The Position of Psychiatry in the Understanding of Human Disease', *Archives of General Psychiatry*, **32**, 1500–1512.
7. Frank, J. D.: 1974, *Persuasion and Healing*, 2nd ed., Schocken Books, New York.
8. Hall, C. S., and Lindzey, G. (eds.): 1970, *Theories of Personality*, 2nd ed., John Wiley and Sons, New York.
9. Jahoda, M.: 1958, *Current Concepts of Positive Mental Health*, Basic Books, New York.
10. Macklin, R.: 1973, 'The Medical Model in Psychoanalysis and Psychotherapy', *Comprehensive Psychiatry*, **14**, 49–70.
11. Macklin, R.: 1972, 'Mental Health and Mental Illness: Some Problems of Definition and Concept Formation', *Philosophy of Science*, **39**, 341–365.
12. Selye, H.: 1956, *The Stress of Life*, McGraw-Hill, New York.
13. Szasz, T. S.: 1961, *The Myth of Mental Illness*, Hoeber-Harper, New York.
14. Wallace, R. K.: 1970, 'Physiological Effects of Transcendental Meditation', *Science*, **167**, 1751–1754.
15. Wallace, R. K. and Benson, H.: 1972, 'The Physiology of Meditation', *Scientific American*, **226**, 84–90.
16. Wolff, H. G.: 1954, *Stress and Disease*, Thomas, Springfield.

SECTION III

PHENOMENOLOGICAL AND SPECULATIVE VIEWS OF MENTAL ILLNESS

J. H. VAN DEN BERG

A METABLETIC-PHILOSOPHICAL EVALUATION OF MENTAL HEALTH

The title of my paper contains terms which will require a brief elucidation. I begin with:

I. MENTAL HEALTH

Seemingly the definition of 'mental health' presents few difficulties. According to a widespread general conception, human existence is a physical as well as a psychological existence, based on the equally well-known division of human existence into body and soul. The body is that part of us which is visible. It is the physician who concerns himself with the health of the body. If the organs of the body are in good condition, which as a rule can be deduced from the usually clear results of scientifically oriented examinations, then a state of physical health appears to be achieved. The soul is said to be that 'thing in us' which is non-visible, which is immaterial. The soul contains all that which is left of us when we remove – in thought – the purely physical elements from our existence. What remains is, for instance, our interests, our longings, our happiness, our sorrow. Our life! Perhaps our destiny. To the domain of our soul belong likewise all the ties we have with others. All these aspects I have just mentioned form together that which is non-physical in the medical sense. It must be noted, however, that the psychic functions here indicated and many others which could be added to this list cannot do without the human body. What is our sorrow if it cannot express itself through our body? What are relationships between people if those people are not physical? But let me stop worrying about this just now, at the very beginning of my paper. Body and soul – it is a well-known and generally accepted distinction. Equally accepted is the distinction between physical and mental health. Today I wish to talk about mental health and not about the functioning of physical organs.

II. PHILOSOPHY

I should not have many difficulties with the second term in the title either. He who studies philosophy, the love of wisdom, as the word implies, is invited

H. T. Engelhardt, Jr. and S. F. Spicker (eds.), Mental Health: Philosophical Perspectives, 121–135.

to treat his subject wisely. That means, that it is my task today to talk about mental health with as much wisdom as is given to me. What it amounts to is that I will attempt to have a look behind the pure facts. He who wishes to talk about mental health in a philosophical way, will try to give a wise, gentle definition of mental health, and that, in my opinion, is only possible if one tries to trace its development in human history.

III. METABLETICS

The word metabletics, which will not be familiar to all of you, is derived from the Greek verb *metaballein*, meaning change. Metabletics deals with cultural changes. In this sense metabletics is a form of history. I admit that this explanation is still very vague. Metabletics as a historical discipline is characterized by some distinct qualities. Primarily, these two: First, that in viewing history it stresses the *human aspect*, the aspect of human everyday life, the aspect of living together. Second, that the historical changes in man's existence are seen as *genuine changes*, and this to a considerably greater extent than is generally accepted in studies of history. Considering the fact that the emphasis in my lecture is on this second point, and in order to make myself clear on this point right now, I would like to give you forthwith an example. This example lies outside the theme of my lecture.

IV. EXAMPLE OF CHANGE IN THE METABLETICAL SENSE

It is well-known that Jean-Jacques Rousseau was the first to write about adolescence as a period of life at which a transition occurs from childhood to adulthood. We find the description of this period of transition, called by Rousseau 'second birth' in his *Emile*, dating from 1762. If we are willing to disregard a few vague references to this period of transition in classical antiquity, we must admit that the short passage in Rousseau's *Emile* is indeed the very first written account of adolescence, of psychic puberty. In general, this was and is still considered as a discovery of a period of life which had always existed. Rousseau would then have discovered adolescence as a botanist discovers a new plant in a hitherto unknown part of the world: the plant has always been there but had not been found up to then. This kind of argumentation is unacceptable in metabletics. Many scholars of fame have long before the time of Rousseau studied the different phases in life. A good example is Montaigne who in his *Essais* (dating from the end of the 16th century) goes as far as to devote an entire separate chapter to the problem of youth and upbringing. Yet not a single word is said about adolescence – he says as little about this psychic growing into adulthood as all the other authors who wrote about growing up prior to Rousseau's time. It is hardly believable that all these writers have been too short-sighted to observe something as much in evidence as this particular period of life. In metabletics it is therefore assumed that adolescence *did not exist* before Rousseau. Now the question of course is: how then did adolescence come into being? The answer to this question and similar questions must also be found within the realm of metabletics. As to adolescence the answer is not difficult to find. Such a distinct phase of life as adolescence can only become imperative when adults have grown *too adult*, that is, when the world of the adult has become too complex for the growing child. With the onset of the natural sciences, the 18th century had indeed become too complex a world for the child. Therefore the child had to have a period of transition so that he eventually might be allowed to enter the world of this new adulthood. At this point you may want to ask a great many questions, but as my short metabletic explanation of Rousseau's first description of adolescence is given by way of an example only, to show how metabletics can consider a new description as an indication of a change in human life, I take the liberty to override these questions.[1]

V. SYNCHRONISM

Metabletics has yet a few other distinguishing qualities which need not be discussed here. There is, however, one characteristic trait which I must not fail to mention. It is the characteristic of synchronism, more a methodological principle really than a characteristic. I have used this methodological principle just a moment ago when I was arriving at my explanation of the origin of adolescence. Society was becoming too complex, adulthood too adult and subsequently adolescence became a necessity. So synchronism makes the rise of psychic puberty or adolescence metabletically intelligible. In what follows I shall again and again employ the methodical principle of synchronism.

VI. HISTORY OF ANTHROPOLOGY

I propose that the title 'a metabletic-philosophical evaluation of mental health' be preceded by what could be described as a-very-short-history of anthropology. I realize that by doing this I embark upon a subject which extends by far the boundaries of this paper. But all I want to do is to point out a few milestones in the history of anthropology.

VII. FIRST MILESTONE: THE BEGINNING OF WESTERN ANATOMY

It seems to me that the first great milestone in the history of the life of Western man is to be found in the year 1306, when the Italian Mondino dei Luzzi (in Latin, Mundinus) opened the body of a dead man, in order to reveal its anatomy, that is, its inner unseen structure. Perhaps in classical antiquity the human body had been opened once or twice for the purpose of an examination of the bodily structure, but in the history of Western European civilization Mundinus was the first to do so. Ten years later, in 1316, his book entitled *Anathomia* (with an *h* indeed: Mundinus made this mistake in the spelling) appeared, in which work Mundinus reports on his findings. This report is disappointing to any person who knows something about human anatomy. What Mundinus was able to see, to discern in that human body was very little, so little that one is inclined to say: nothing. But more could hardly be expected. It was after all quite revolutionary as it is to take a knife and cut up a body. His disciple, Guido da Vigevano likewise saw next to nothing inside the human body, as is proved by his anatomical treatise, printed in 1345. I would have no reason to mention this second work, nor its author, were it not for the fact that Vigevano included in his anatomy a number of illustrations – and those are missing in the work of Mundinus. These illustrations *show us* that the anatomists of the first hour not only did not but *could not* distinguish much in the human body. The first picture in Vigevano's work exhibits the anatomist making an incision in the human body (see Figure 1). You can see Vigevano, the physician and anatomist, and next to him (standing upright) the dead man whose body Vigevano opens up with a knife. In one way or other they must have fixed the corpse to a board behind him. I will ask you to pay attention to Vigevano's eyes. One would expect Vigevano to look at the knife. But he does not. He looks at the closed eyes of the dead man. He has a relationship with the corpse, with the man of that corpse, with this deceased fellow-human being. He puts his left hand in an affectionate, intimate manner around the dead man's body. What this picture, dating from 1345, reveals, is the relationship which the physician has with his deceased patient – a relationship which in a sense is terminated by the knife.

VIII. SECOND MILESTONE: THE ANATOMY OF VESALIUS

More than two centuries elapsed before a true 'modern,' and in this sense splendid textbook of anatomy saw the light, *De humani corporis fabrica*

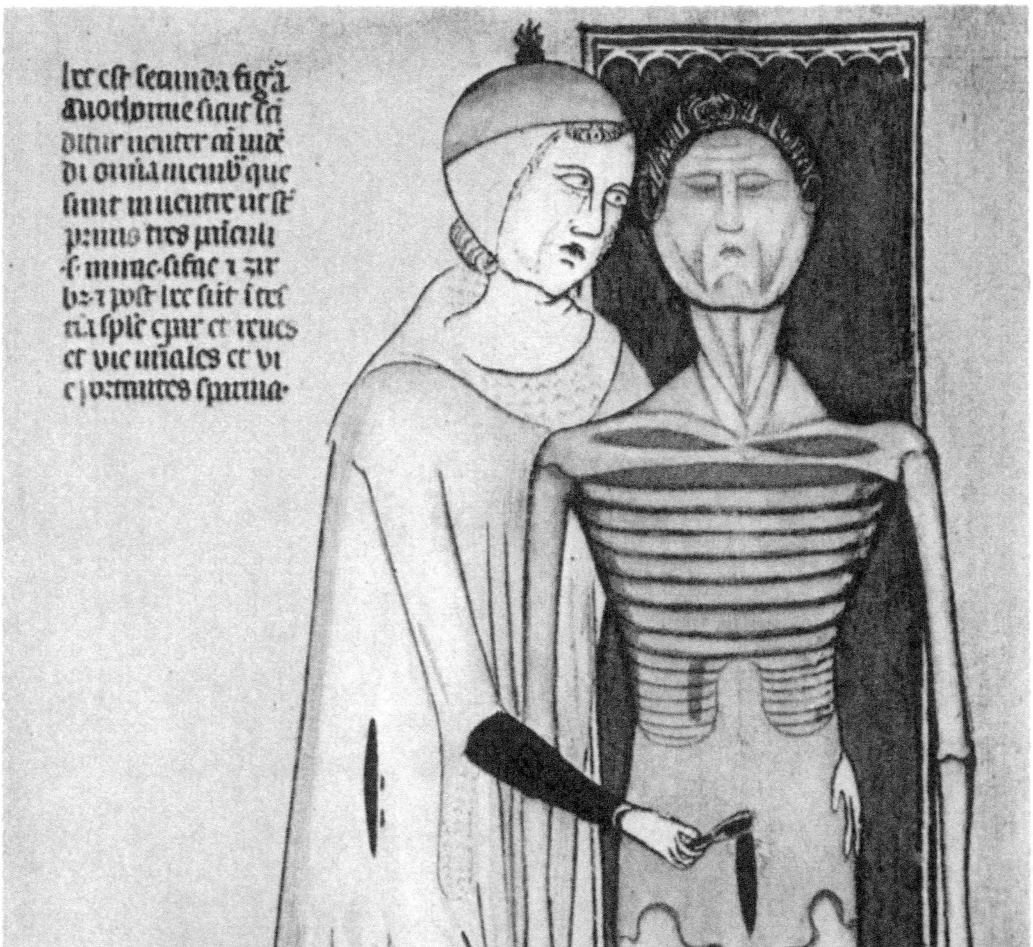

Fig. 1. Vigevano, 1345

Fig. 2. Vesalius, 1543

written by Andreas Vesalius and published in 1543. In this book we find described and illustrated practically everything of macroscopic human anatomy. In this drawing too (see Figure 2) you can see the anatomist, Vesalius in person, and a corpse. Vesalius is not looking at the body. He is looking *at us*. His eyes seem to invite us to take a look at, better *into* the dissected body. Vesalius demonstrates an arm with a number of well displayed anatomical details. His left hand is curved round *this arm*: it is curved round an anatomical preparation. The corpse itself is only partly shown. The corpse is no longer of interest. Neither does Vesalius seek any contact with the dead man. He expelled the dead person from the dead man's body and he, the anatomist now triumphs over the dead man's body. He, Vesalius, invites us to look at this object: into the body, into that *thing* separated from man.

Now what is the meaning that these two pictures, the one by Vigevano and the other by Vesalius, want to convey? They want to tell us that man has been alienated from his own body. In other words: that man has changed, has become two, has been divided into two parts: body and soul, and that, this is important, the body has nothing to do with actual human existence. Let me put it like this: the origin of the differentiation between physical and mental health can be found in the years stretching from 1300 to 1543. This differentiation did not exist originally: it is a 'made' differentiation, an *artefact*.

IX. THIRD MILESTONE: THE HUMAN HEART IS A PUMP

The third milestone stands in the year 1628, when William Harvey in his little book *Exercitatio anatomica de motu cordis et sanguinis* proved that the human heart is a hollow muscle, a pump as he said. The heart had never been anything like that before. It was supposed to be the seat of feeling, of passion, of faith. Harvey put an end to this belief. He deprived this already estranged body of the heart, of this living, feeling, relating passionate centre.

X. FOURTH MILESTONE

The fourth milestone can, to my belief, be identified in the year 1733, which year saw the publication of George Cheyne's *The English Malady*. It is in this work that we find the first description of what later at the end of the 18th century was to be called a *neurosis*. The ancient classicists never wrote about neuroses. True enough, people knew about hysteria, but hysteria is a constitutional disturbance and not a genuine neurosis. There were more constitutional disorders which were erroneously called neuroses. Genuine neuroses, in the sense of psychic disturbances of the (in essence) mentally healthy person, were first described by the Scot, George Cheyne, in 1733, as a new disturbance, a new malady, called by him *The English Malady*. You will understand that in metabletics the supposition is made that before the 18th century neuroses actually did not exist. This is indeed my conviction. In any event, it is of considerable importance to the history of anthropology that it was the year 1733 which witnessed the very first description of the neuroses.

XI. FIFTH MILESTONE: MESMER'S 'MAGNETIC SLEEP' AND JEAN PAUL RICHTER'S 'DOUBLES'

The fifth milestone is to be found shortly before and during the French Revolution. It would actually be more appropriate to call it a 'small bundle of milestones.' First, around 1780, Franz Anton Mesmer discovered that with what he believed to be his magnetic power, some of his patients got into an abnormal state of mind, a condition which was neither one of being awake nor of being asleep. This state was called the 'magnetic sleep' and the significance it has for the psychotherapist is that the state of unconsciousness which some of Mesmer's patients experienced, was similar to that of Freud's first patients. Independently of Mesmer, Armand de Puységur made the same discovery in 1784, when some of his patients manifested a so-called 'spontaneous somnambulism.' The patients who displayed this behaviour would walk about in a manner somewhat unreal and dazed and were not able to communicate with anyone but Puygésur himself.

Straightway I must put on record here that in 1791 the German writer Ludwig Tieck, in his short story *Ryno* gave the first description of what later in literature was to be called the *double*. The man who actually used the word *double* – *Doppelgänger* – for the first time, was Jean Paul Richter, in 1786. Since then the *Doppelgänger* has never left the pages of Western literature. I believe that human existence has since the end of the 18th century *turned into a double existence*, which double existence was already revealed by Mesmer's and Puységur's patients. In 1800 Goethe writes: '*Zwei Seelen wohnen, ach, in meiner Brust*' (Two souls, alas, dwell in my breast), and this, I believe, is precisely what happened at the end of the 18th century. Human existence had split into two.

XII. SIXTH MILESTONE: THE ALMIGHTY UNCONSCIOUS

The sixth milestone is set in place about 1900. In 1900 Freud published *The Interpretation of Dreams* (*Die Traumdeutung*) and subsequently, in 1901,

Psychopathology of Every-Day Life (*Psychopathologie des Alltagslebens*). These two works show that the unconscious, so cautiously coming to the fore at the close of the 18th century, became, one could say, *almighty* around 1900. It dominates not only the sleeping but also the waking hours of life. *Consciousness*, in Freud's opinion, was no more than a 'surface phenomenon,' as he literally said; all the rest, the 'great bulk' of our personality was *un*consciousness. Hereby it must be stated with emphasis that the almighty unconscious was considered to be a kind of *anti-ego*, that is, a *pathological entity*. This was exactly the case with the introduction of the motion of the unconscious at the end of the 18th century. It could be maintained, that around 1900, man was being overpowered, as it were, by his *pathological Doppelgänger* – who came into being as a dwarf shortly before 1800.

XIII. SEVENTH MILESTONE: WANING OF THE UNCONSCIOUS

The seventh milestone is prompted by the fact that about 1945 two publications appeared in which the psychotherapist is beginning to doubt the existence of 'the unconscious.' The first publication is by Carl Rogers who in 1942 published his *Counseling and Psychotherapy*, in which little emphasis is laid upon *the unconscious*, and this signals (in a book dealing with psychopathology) an extraordinary fact. The second publication in which the existence of *the* unconscious is actually denied, is by H. S. Sullivan: it is his article entitled *The Meaning of Anxiety in Psychiatry and in Life*. Sullivan refuses to use the word unconsciousness, he talks about the '*unnoticed*,' the '*unaware*.' Since then an increasing number of psychotherapists have come to the fore who tend to repudiate the existence of that which was called 'the unconscious.' Neurotic disturbance has been, since 1945, less and less a conflict between two parts of the subject, and increasingly a conflict between concrete persons. In other words: *neurosis became an interpersonal instead of an intrapsychic phenomenon*. Consequently, neurosis is henceforth to constitute an aspect of *normal* human living. Since 1945 neurosis has ceased to be strictly pathologic. In reverse, it would be possible to contend that normal human existence and/or normal human relations have become pathological since about 1945.

XIV. SEVEN MILESTONES

I have marked seven milestones in the history of anthropology. The first three relate to the human body; the result was the separation of body and

Fig. 3. St. Francis Preaching to the Birds. Giotto, shortly after 1300.

Fig. 4. The Large Piece of Turf, Albrecht Dürer, 1503 (Vienna, Albertina).

mind. After William Harvey had called the heart a pump, the dualism of body and soul was stressed further by Descartes. The publication of the first book on neurosis marks the fourth milestone, the milestone of the middle; the soul severed from the body is beginning to show signs of distress. The three consecutive milestones demonstrate how this distress revealed itself as a split in two, of which the unconscious part was initially feeble but then became all-powerful and subsequently, in our days, has changed into a normal phenomenon, manifesting itself as a characteristic of normal human relations.

What does this development mean? This question is the essence of my paper. If I want to apply the methodical principle of synchronism in answering this question, I have to reformulate my question: can we point to other facts that occurred in the small epochs marked by the seven milestones? A summary of events is easily given.

XV. EMANCIPATION OF THE LANDSCAPE

Between 1300 and 1350, that is, in the days of Mundinus and Vigevano, painters like Giotto (see Figure 3) and the Lorenzetti brothers started to paint the landscape as something existing by itself, that is, apart from man. The *complete* emancipation of the landscape was achieved later, in the period between 1500 and 1550, and can be seen on the canvasses of Albrecht Dürer (see Figure 4) and Altdorfer, who were the first to paint a landscape as a *separate* entity, that is, *without man* featuring in it. The years between 1500 and 1550 were the period of Vesalius. Summing up: simultaneously with the development of Western anatomy and in the wake of this development of the separation of body and soul, we witness a separation of man and his world in the paintings of that period. In the paintings, yes! But painters never paint an important new fact on their canvasses if life itself has not acquired that new fact.

XVI. THE PHYSIOLOGICAL HEART AND THE SACRED HEART

In 1628, the year of the next milestone, William Harvey declared that the heart is a pump, a physiological organ. One year later, in 1629, Jean Eudes writes a letter to Madame de Budos, abbess at Caen, in which he mentions for the first time what later would be called the Devotion to the Sacred Heart. Jean Eudes is the founding father of the Sacred Heart Devotion. This amounts to the fact that at the time Harvey called the heart a *pump* and in this way *degraded* the heart, Jean Eudes was trying to *exalt* the heart and raise it to the lofty position of a *sacred heart*. According to the Sacred Heart Devotion, the human heart, if it is a devout heart, becomes one with the heart of Christ. I cannot see in the work of Jean Eudes more than an act of sheer desperation. The human heart in that body separated from man was a lost cause. In desperation Eudes wanted to lift the heart to the rank of a sacred heart.

XVII. THE MILESTONE IN THE MIDDLE

We now come to the milestone in the middle: George Cheyne, with his first description of neurosis in *The English Malady*. That was in 1733. In the selfsame year, John Kay of Bury, Lancashire, constructed the first little machine of the Industrial Revolution, his 'flying shuttle.' That is to say: simultaneously with the first description of neurosis we witness the first symptoms of what was called the Industrial Revolution. One could put it like this: by the time the emancipation of the landscape was completed, the first little machine *was set up in that landscape*, the Industrial Revolution could take place in that landscape, the world could be prepared for industrialization. In other words: into this world, where man had been estranged from his surroundings, a further element of alienation was introduced: industrialization. Is it so curious then that neuroses should arise? Neuroses can be regarded as the pathological consequence of a world which has been steadily turning away from man.

XVIII. THE END OF THE OLD ELEMENTS

We now take a look at the last three milestones in order to see which are the synchronic facts that accompany them. These were the last three milestones in anthropology. First: shortly before 1800 the birth of the double as a manifestation of the pathological unconscious. Second: about 1900 the rise of the almighty, equally pathological unconscious. Third: around 1945, the waning of the unconscious, the end of the pathology of neurosis, and the beginning of 'neurotic normal life.'

The synchronisms will not raise many difficulties if only we remember that man's landscape was set free in the years between 1300 and 1600.

In 1784 Antoine Laurent Lavoisier established that water was a compound of oxygen and hydrogen. Before Lavoisier water was generally considered to be an *element*, next to the other three elements. fire, air, and earth. Since 1784 the four old elements, which after all constituted four essential substances of our universe, have been replaced by a host of novel, anonymous elements: the elements of the chemistry researcher.

In and around 1900, the next milestone, it appeared that these new elements could also be subjected to a division. The atoms can be split. Simultaneously, and most certainly closely linked to this discovery, was the discovery by Einstein in 1905 of the mass–energy equation. At this point you will undoubtedly be able to guess the synchronic facts of 1945: it was the year 1945 that witnessed the first atom bomb explosion.

Summing up: the last three milestones are the milestones of the pathological subject. Concomitant with the pathological subject went the disintegration of matter. The moment, and that is around 1945, that psychological pathology becomes part of normal human existence, we enter the age of atomic explosions.

XIX. SUMMARY

I have now come to the end of my exposition, which concerned itself with mental health. I fully realize that there is more than one way of dealing with this subject and I am conscious of the fact that a completely acceptable representation of the history of mental health can be given in a manner different from the one adopted by me. I wanted to give you an unusual view of the history of the concept of mental health: metabletic interpretation – 'metabletic-philosophical evaluation,' as the title of my paper reads. It was no more than an attempt to trace, by means of the method of synchronism, the events which accompanied the development of anthropology, of sane and sick human existence, and, in particular, that of undisturbed and disturbed mental health. The historical parallels which have been found have such marked correspondence with the dates of important events in the history of anthropology, that I myself cannot doubt the existence of an essential link. So I came to the following conclusion.

In the course of centuries human existence has in a very specific sense been impoverished. To begin with, we witness a steadily growing separation of man and the universe. Synchronically there is an equally steady growth in the separation of mind and body. The human subject is being driven into a corner which is getting smaller and smaller and there, in that little corner, begins the pathology of the human subject. Then the old elements are chased out of the universe, and when in this universe, which has been severed from man, the new elements appear, their atoms are split into tiny, the tiniest particles, and pathology threatens to overwhelm the subject. Today we know that *our relations* are ruled by the abnormal, and this is probably even worse. Is it by accident that we live in a world that has enough atom bombs at its disposal to stop all life on earth?

Now, what is the history of mental health? Without doubt it is the history of the personal health of the human being. But it is at the same time the history of *matter*, the development of the pathology *of the plane upon which we live*, of the growth of *the disease of the elements* which make up our earth.

Mental health: *the expression is wrong*. There is no such thing as *mental health*. There is *health*, and this health encompasses everything that exists: *health of the unity of man and his landscape, man and his body*. The separation of body and soul, and that other separation of man and world are expressions of the fact that we fall short of that health.

You asked me to give a metabletic-philosophical evaluation of mental health. Looking back to the milestones in anthropology and to the events that synchronically can be distinguished, I come to the conclusion: we would never be talking about mental health if we had not already for some hundred years been suffering from a general and in some ways gentle disease that encompasses our world. The first, most important symptom that there is something wrong with our *mental health* lies in the word 'mental.' Yes, if we are asked to live *mentally*, that is, as a *soul*, in a strange anatomical body, in a strange chemical–physical world, nobody can expect us to live in good health.

Institute of Conflictpsychology,
University of Leiden, Leiden, Holland

NOTE

[1] Readers wishing to learn more about the 'metabletic' way of thinking may consult J. H. van den Berg, *The Changing Nature of Man*, Dell, New York, and *Divided Existence and Complex Society*, Duquesne University Press, Pittsburgh.

RICHARD M. ZANER

SYNCHRONISM AND THERAPY

My role as commentator I understand to involve me in a double responsibility: to accept the invitation intrinsic to every epistemic endeavor, first to 'think-along-with' the author (*co-mentari*), in order to understand what is said and maintained, and, second, to test it critically (i.e., again, dialogically). I want first, then, to lay out what I take to be the main themes of Prof. van den Berg's presentation, and then to raise or mark out some issues which call for our closer scrutiny and reflection.

I

'Metabletics deals with cultural changes. In this sense metabletics is a form of history . . . a historical discipline' ([4], p. 122). More particularly, this discipline (1) stresses the *human aspect* in history, and (2) argues that historical changes in human life are not ephemeral variations on an underlying, unchanging human 'nature' but are rather essential changes ([4], p. 122). The 'methodical principle' ([4], p. 123) which enables us to make the latter intelligible is 'synchronism' – which is not only a matter of 'method' but is a quality intrinsic to human history itself. Indeed, as I understand him, Prof. van den Berg's point is that *historical method* itself is grounded in *human historical life* (in the historicity of human existence): it is because the latter displays synchronism that the former is available to us and synchronism is made necessary as a method.

Thus the 'method of synchronism' makes it salient that there is an 'essential link' ([4], p. 134) between events central to the history of anthropology and other, temporally concurrent events in art, science, technology, or culture more generally. At the same time, it should be emphasized that metabletics also purports, as I see it, to educe, explicate, or make prominent that there are essential links among historically sequential periods of human history. Thus, we might say, *dia*-chronics has equal status with *syn*-chronics. Just this is what is supposed to have been marked out with the example of adolescence ([4], pp. 122–123), as also of course the central theme of the presentation, the 'milestones' which are internally *synchronal* and, across historical periods, *diachronal*. As adolescence authentically *emerged* (e.g., was *not* present in human culture prior to the *complication* of adulthood initiated

H. T. Engelhardt, Jr. and S. F. Spicker (eds.), Mental Health: Philosophical Perspectives, 137–142.

during the time of Rousseau – which saw not only that, but also such changes as affected the nature of the house, its internal spacing and correlated functions) so are each of the milestones historical emergents, genuine changes in human existence itself.

Necessarily able to provide only a sketch of these profoundly significant occurrences in (Western) history, Prof. van den Berg maintains that dating from 1306 (when Mundinus first cut into a human cadaver), and continuing in a sequence of decisive events to the present, 'human existence has in a very specific sense been impoverished' ([4], p. 134). Deeply correlated with that intrusion into the human body – which 'synchronously' (and diachronously) had the force of initiating the separation of mind and body, and thereby introducing the cultural sub-texture for the body-and-soul dualism which gave credence to and was definitively strengthened by the Cartesian ontological bifurcation – there becomes evident an elemental 'distress' affecting the soul itself, an internal schism. Scarcely a century after Descartes, indeed, this schism shows up and is formally termed 'neurosis' by George Cheyne (1733); and, a few decades later, the abnormal state of mind termed 'magnetic sleep' by Mesmer (1780) and the 'double' by Richter (1786) is even more prominent. (Indeed, not long afterward, and prior to the advent of Freud, Kierkegaard was able to lay out the incredibly complex entanglements of that decisive disease of the soul, despair (1843).) And, after Freud's attention was drawn so strongly to the internal divisions of the psyche, by 1945 this psychic pathology has become part of normal human existence – in the age of the atom.

This historical schisming of the human context is, he argues, synchronously matched in other crucial cultural events: e.g., in painting with the 'emancipation of the landscape.' The 'circum-stance' of human life is split off from human life, and thus is the soil prepared for and the germs of estrangement planted in that context. Within this setting there could now occur the initial cultural form of alienation – industrialization ([4], p. 133). Given this, it is little wonder that neurosis could make its first appearance: 'Neuroses can be regarded as the pathological consequence of a world which has been steadily turning away from man' ([4], p. 133). And, today, the splitting of the atom, and the consequent cultural presence of 'explosiveness,' metaphors the condition of man: the objectifying of internal schisms, the splitting apart of man from man, of man from himself. Along with the development of the pathological subject goes the disintegration of matter ([4], p. 134).

Thanks to this metabletic uncovering of historical *anthro-pathology*, the entire issue of 'mental health' receives a new significance. 'There is no such thing as *mental health*. There is *health*, and this health encompasses everything that exists: *health of the unity of man and his landscape, man and his body'* ([4], p. 135). This 'health', of course, we necessarily only poorly approximate today. As Gabriel Marcel once pointed out [2], we would never witness so much frenetic talk of and search for the 'person' and 'integrity,' e.g., in our times, if we had not historically witnessed their disappearance as *effective realities* – as realities to be reckoned with. Or, as Prof. van den Berg concludes:

the most important symptom that there is something wrong with our *mental health* lies in the word *mental*. Yes, if we are asked to live *mentally*, that is as a *soul*, in a strange anatomical body, in a strange chemical–physical world, nobody can expect us to live mentally in good health ([4], p. 135).

'Health' – that is, that integral 'heal-ing' manifested solely by genuine 'wholeness' [1] or the living integrity of the total contexture, the embodied-person-within-the-full-fleshed-*Umwelt* [5] – is thus unavoidably, in our times, a faint, wispy trace of the wholeness which has been historically split apart.

II

Prof. van den Berg's essay is a remarkable one - in many ways a tour-de-force of sweeping reach. And, I must say, his conception of metabletics, with its double-focused syn- and dia-chronism, is not without its dramatic range, its sensitive and *ironic play* – the specific scope of which I myself am quite anxious to bring into relief, to tie down securely on pains of having his intriguing creature of metabletics, like the statues by ancient Daedalus, up and out before we realize what has happened. This effort has its risks, not unlike the efforts, in 1492, to save Pope Innocent VIII: in a deep coma, an effort to save the elderly pontiff was undertaken, by pouring down his throat the blood of three healthy young men. It was no use, of course: he died immediately, and so did the donors. One can easily kill the 'play' and the 'irony,' and in many ways, and efforts to save them should be cautious indeed not to blunder it about unconscionably.

First, then, to prepare the soil. Nothing so uniquely characterizes a culture as the image of man it fabricates. The image, if you will, is a potently reflexive metaphor of itself, revealing of the subtle textures of that culture, and manifesting itself in varying ways throughout. Thus, as Prof. van den Berg says, those historical parallels he finds 'have such marked correspondence with the dates of important events in the history of anthropology,' that he 'cannot doubt the existence of an essential link' ([4], p. 134).

But, 'synchronously,' *our times are surely not an exception to this*. To paraphrase Max Scheler, however [3], whose view is close to that of Prof.

van den Berg, never in recorded history has man been so problematic to himself as in our own. Our image of man is fractured; or better, our cultural repertoire is riddled by a multiplicity of deeply conflicting theological, philosophical, and scientific images which seem to encounter one another only as antagonists. Precisely this fact, indeed, lies at the heart of the oftentimes exquisite and painful dilemmas in contemporary medical practice and theory. What we must realize, Prof. van den Berg insists, is that this schism is merely part of the story: not only has the human psyche been splintered, along with its embodying organism, but matter itself, '*the planet upon which we live*' ([4], p. 135) has been split apart. 'Concomitant with the pathological subject went the disintegration of matter. The moment . . . that psychological pathology becomes part of normal human existence, we enter the age of atomic explosions' ([4], p. 134).

(Merely to suggest how powerful is this conception, one could I think readily correlate signal events in the history of philosophy with those which he has sketched for us – from the times of Cusa and later Galileo, through Hobbes and Descartes and Pascal, up to Hume and Kierkegaard . . . .)

In short, we are obliged to contend, not merely with one among many other, equally legitimate portrayals of the history of mental health, not with a merely 'unusual view' ([4], p. 134) of it, but with an extremely fruitful and consequential theoretical vision of human-being-in-the-world. It is one which says: the history of human existence is one which manifestly gives the lie to the idea of an unchanging human *nature*: human existence is found in its *historicity*. Indeed, the very idea of 'nature-izing' human reality is itself an historical consequence of *genuine change* in human reality. As Husserl had remarked in 1910, 'naturalism' with its naturalization of consciousness, is a consequence of the discovery of nature.

Now, however, several issues must be marked out.

(1) If I have understood him, there are several matters – 'matters' which, if you will, *matter* to us – that must be pointed to. They are integral to the 'play' and the sensitive irony of the conception. If it is *not* 'by accident' that the normalization of psychic pathology in our times occurs synchronously with the advent of the splitting of the atom – *neither can it be accidental that the kind of conception Prof. van den Berg himself presents should have come when and as it has*! Without detracting from its own originality in the least, the view is nonetheless closely tied to its own history – in philosophy, at least, one finds this posture in the works of thinkers such as Pascal, Kierkegaard and Nietzsche, through Brentano, Dilthey and James, to Husserl, Heidegger, Erwin Straus and others.

My point is this: *what is here called metabletics is itself synchronously and diachronously within the self-same history which on the other hand it has marked out for us*. That is: the historical consequence of the history of anthropology cannot, *in terms of metabletics itself*, be only the momentous tripartite pathology he gives us (psyche, body, world), for the metabletical effort to explicate that complex pathology and (as I shall presently argue) to *treat* it is itself emergent from that very history. In still different terms: the history of mental health teaches us our own curiously cunning and ironic condition, that along with, and deeply tied to, the schisms he points to is the increasing presencing of a *self-conscious* (reflexive) *historicity*, an equally potent tendency to 'draw together' (if you will, not merely to syn-thesize, but to syn-chronize reflexively).

To try to get out of such linguistic acrobatics, the echoing baffles of such words: we must not forget that not only is Prof. van den Berg, but so are we all, integrally of a piece with the very cultural milieu he describes. That history is *our* history. Metabletics, along with the historical reflections of a Dilthey, the phenomenology of Husserl, the ontology of Heidegger, the psychoanalytics of Sullivan and Rogers, the hermeneutics of Gadamer and Ricoeur – all are themselves emergent with the very context which has yielded the normalization of pathology. None of these medical, psychiatric or philosophical inquiries is outside the historical development analyzed by Prof. van den Berg. We, as readers of the paper, then, are in effect invited to apprehend ourselves in that history – which is our own. And just this constitutes both the play and the profound irony of the paper – in part!

(2) The other part is equally complex, cunning and entangled, though I must be briefer in trying to tease it out. There are two facets of the point. (a) On the one hand, while metabletics is presumably an 'historical discipline,' seeking to explicate or educe the synchronisms of the history of anthropology, that is not all it is – or so it seems to me. Though I may be skating on thin ice here, I detect a hidden rhetoric, one whose force derives from a profoundly *ethical* concern. Making us cognizant of *real changes* in human existence, this analysis seems as well to make us desire to 'do something' to correct the 'wrong' of those changes. Thus, when Prof. van den Berg writes: 'Today we know that *our relations* are ruled by the abnormal,' this is not an axiologically innocent historical description; for he goes on immediately to remark: 'and this is probably even worse' ([4], p. 134). In other words, unless I miss my mark, metabletics is a profoundly *therapeutic* discipline, one showing marked rhetorical devices used in the service of an ethical concern for the splits and schisms it delineates.

(b) And, as with any effort to serve, take care of, cure – which is not only descriptive of changes but which *seeks to bring change about* – not only is there the clear ethical concern, but as well an idea of the 'for the sake of,' the goal, toward which the therapy is necessarily oriented. Metabletics in other words proceeds on the assumption that it has after all discovered something which *has always existed* – not in the sense of a 'nature' but in the sense of what I may call the *essential condition of human existence*, or of its *authenticity*. This is found explicitly in Prof. van den Berg's presentation at the very end, and concerns the essential meaning of 'health:' '*health of the unity of man and his landscape, man and his body*' ([4], p 135). Expressive of the authentic, essential significance of human existence, the historical separations of man from his body, and of man from his landscape, 'are expressions of the fact that we fall short of that health' ([4], p. 135).

Given this conception of the 'unity,' or as I should prefer to say, the '*contexture*,' of man-soma-world, as 'health' ('wholeness' or 'healed'), and thus as what is authentically essential to human existence, what I have called the therapeutic character of metabletics seems quite clear. And, with this, the marvelous 'play' and irony of Prof. van den Berg's presentation can be seen in its richer sense: metabletics is, as historical syn- and dia-chronics, *emergent from the very history it delineates, and is emergent precisely as a 'heal-ing' discipline*. What must be 'healed,' however, that is, human existence itself, has become not only fragmented but exceedingly cunning, shifting, baffling and baffled, convoluted and reflexively acrobatic. And thus must this discipline focused on human existence itself be a discipline of dialectic skill, artful and sly in its effort to entrap, even ambush, its patient into unconcealment.

Southern Methodist University,
Dallas, Texas

BIBLIOGRAPHY

1. Kass, Leon: 1975, 'Regarding the End of Medicine and the Pursuit of Health', *The Public Interest* **40**, 27, 28–29.
2. Marcel, G.: 1952, *Man Against Humanity*, trans. by D. Mackinnon, Harvill Press, London.
3. Scheler, Max: 1961, *Man's Place in Nature*, trans. by H. Meyerhoff, Beacon Press, Boston.
4. van den Berg, J. H.: 1977, 'A Metabletic-Philosophical Evaluation of Mental Health', in this volume, pp. 121–135.
5. Zaner, Richard M.: 1975, 'Context and Reflexivity: The Genealogy of Self', in H. T. Engelhardt, Jr. and S. F. Spicker (eds.), *Evaluation and Explanation in the Biomedical Sciences*, D. Reidel Publishing Company, Dordrecht, Holland, pp. 153–174.

EDITORS' NOTE

This session, pertinent to the theme 'Mental Health: Philosophical Perspectives,' is dedicated to the memory of Professor Dr. medicine Erwin W. Straus, who died on May 20, 1975, at Lexington, Kentucky, the home of the Lexington Conferences on Phenomenology: Pure and Applied, all of which he saw through the press subsequent to his 73rd birthday. He was in his 84th year at the time of his death, and notwithstanding his longevity, his death was still untimely.

In 1970 he published 'The Miser,' and now, in retrospect, we realize that many of us still want to possess him and to keep him before us. Yet should we do so, distracted from our work, he would, were he here, chastise us (those who knew and loved him) for risking ourselves at the edge of the pathological. Ours has to be a healthy bereavement. On what better occasion than this, given our theme, should we acknowledge our debt to Erwin Straus and the legacy he has left for all thinkers who seriously attend to the human condition. To register that legacy we have appended a Bibliography of Erwin Straus' writings.

H. T. Engelhardt, Jr. and S. F. Spicker (eds.), Mental Health: Philosophical Perspectives, 143–155.

STUART F. SPICKER

THE PSYCHIATRIST AS PHILOSOPHER[1] GEWIDMET ERWIN W. STRAUS, M.D., PH.D. (h.c.), LL.D. (h.c.) (OCTOBER 11, 1891–MAY 20, 1975)

I. INTRODUCTION

In the Summer of 1965, some months after I first met Erwin Straus and his confidant and collaborator, the late Richard Griffith, I was privileged to participate and work in their research setting in Lexington, Kentucky, a city now celebrated as the home of the Lexington Conferences on Pure and Applied Phenomenology, the fifth of which convened in 1972.[2] On the occasion of my visit I began my own compilation and personal study of Professor Straus' complete writings,[3] beginning with 'Zur Pathogenese des chronischen Morphinismus' [10], his inaugural dissertation, published in 1919 upon attaining his doctorate in medicine at the Friedrich Wilhelms University in Berlin.

As was the case with innumerable others, I was especially captivated by the now classic 'Die aufrechte Haltung: Eine anthropologische Studie,' which made its first appearance in 1949 in the *Monatsschrift für Psychiatrie und Neurologie* [12], a *Festschrift* celebrating the 80th birthday of Karl Bonhoeffer, professor of psychiatry and father of the theologian Dietrich. This significant paper has made a series of reappearances: It was extensively reworked and expanded, appearing in English in 1952 in *Psychiatric Quarterly* [12]; the German re-emerged in 1960 in *Psychologie der menschlichen Welt* [12], Straus' first *Gesammelte Schriften;* the English version was republished in 1966 in his second volume of selected papers, entitled *Phenomenological Psychology* [12]. In 1971, upon the 450th anniversary of the founding of Straus' Gymnasium in Frankfurt a/Main, the editors of *Jenseits von Resignation und Illusion* once again selected the German version for their collection [12]. In 1973 the English version appeared once again in an anthology edited by Richard Zaner and Don Ihde [12].

Is it possible that notwithstanding six separate appearances of this classic paper, spanning some twenty-four years, its critical philosophical import is yet to be distilled? This bold suggestion is perhaps made even more so by the fact that in 1963 Straus published 'Zum Sehen Geboren, Zum Schauen Bestellt: Betrachtungen zur "Aufrechten Haltung"' in *Werden und Handeln* [17], dedicating it to Viktor E. Freiherr von Gebsattel on the occasion of his 80th birthday. Not only man's upright posture but his 'beholding' (as contrasted with his 'seeing') gaze is brought to bear on his apprehension of the world and all things in it. Surely one might expect that by 1965 with the appearance of Erling Eng's English translation, 'Born to See, Bound to Behold: Reflections on the Function of Upright Posture in the Esthetic Attitude' [17], Straus' total shattering of *Erkenntnistheorie,* Cartesian dualism, and transcendental philosophy should have been noticed by the philosophical and medical community.

Yet in spite of this paper's originality, which even surpassed, in places, the initial version, its import for philosophical reflection (I shall maintain) has not been made salient. There are a few reasons for this state of affairs: First, Straus does not philosophize in what is now the most fashionable philosophical mode of discourse; secondly, psychiatrists and neurologists have not frequently drawn upon his intellectual insights; thirdly, the phenomenological movement, in which he is frequently located, has had enough difficulty gaining the attention and respect of Anglo-American philosophers and has not particularly attracted the eye of philosophers who could most appreciate his new turning; finally, many of us are too prone to view an original thinker in terms of his intellectual indebtedness to others, thereby underplaying his contribution. In my view the crucial philosophical import of Straus' thought is not determined by the development of the *Daseinsanalyse* of Ludwig Binswanger [8, 13], the philosophical anthropology of Merleau-Ponty, or even Edmund Husserl's descriptive psychology and transcendental phenomenology [2]. Binswanger, Merleau-Ponty, Husserl, Heidegger, and others are frequently cited as precursors to Straus' position, the authors frequently forgetting, for example, that his *Von Sinn Der Sinne* [11] made its original appearance in 1935, a decade before Merleau-Ponty's *Phénoménologie de la perception* [7].

Before developing my thesis I hasten to add that in addition to some of the authors included in the *Festschrift* [21], published on the occasion of Straus' 75th birthday, two other philosophers have given testimony to Straus' contributions; if there are others in addition to those of Marjorie Grene [1] and Herbert Spiegelberg [9], I am unfamiliar with them. In their commentaries Spiegelberg remarks that Straus' rôle 'has been that of inspirer' ([9], p. 279), and Grene observes that Straus' work serves as a *'Wegweiser'* ([1], p. 185, also see p. 194). These judgments are surely correct and corroborated by the testimony of many of his students. Speaking personally, I have been and continue to be inspired by Straus' genius and have followed the signpost marked out by his publications and engaging seminars. But in what I am about to say I am aware of the trap that yawns for the disciple who, accepting the master's guidance in perfect devotion, lands himself in an extreme position. I trust that my remarks will not strike others as too extreme, but I believe that notwithstanding Straus' general rejection of the philosophy of Immanuel Kant, he is correctly viewed as taking a special road from Königsberg.

II. STRAUS' CRITIQUE OF TRANSCENDENTAL PHILOSOPHY

Straus rarely acknowledges Kant because, in part, he is suspicious of over-intellectual models. For this same reason he has avoided the Husserlian highroad of transcendental subjectivity (which does not in itself justify Grene's claim that Straus' starting point most resembles that of Heidegger and not that of Husserl). Rather one should remain with Kant as the Strausian analogue. Although Straus rejects Husserl's method of trancendental *epoché* and subsequent reduction, since he always maintained that one cannot discover an absolute consciousness, one should not conclude, as Spiegelberg does, that Straus has no interest in constitutive analyses ([9], p. 268). Rather, he reemploys the notion of 'constitution' but rejects Husserl's and Kant's suggestion that consciousness, or the categories of the understanding, constitute the meaning of the world including others as well, perhaps, as one's own body. It is neatly stated when Straus asks (as he once did of me): 'How can Husserl say *Bewusstseinsleben*[4] when he performs the *epoché* with respect to "world" and "lived-body?" How can Husserl speak of "conscious life" [*Bewussteinsleben* or *vie de la conscience*] when he has suspended, as part of his methodology, the lived-body [*Leib*] from the earliest phases of his philosophical method?'[5] It is precisely at this point that Straus could accept transcendental philosophy no longer. For Straus holds that 'human physique reveals human nature' ([19], p. 164). He maintains that philosophers must begin and remain with man considered in his concreteness. And so, on another occasion, he asked me, rhetorically: 'Does transcendental consciousness sleep?' He rejects, therefore, Kant's and Husserl's archaeology of consciousness and the forms of sensibility and the understanding which, in Kant, are the home of objects as known by us ([4], p. 127, A94–B127). For him, it is rather in and through the structures of our lived-bodily condition that we *constitute* the objects we come to apprehend and of which we come to have knowledge. The most important of these categories is man's *uprightness* and placing of himself in opposition to the all-embracing totality [the Allon] ([16], especially pp. 42–58). But this is not the only category Straus' has commended to us; there are a few others: Once upright, so to speak, we constitute the visible in its splendor by our 'beholding gaze;' furthermore, we constitute the unity of the world, its stabilization and permanence, in terms of our own bodies which (as Merleau-Ponty once remarked) like the artist, we always take with us.[6] Furthermore (and I am intentionally sketchy here) the asymmetry or sidedness of *ánthrōpos* enables the constituting of lived-space in its directionality. In time we even transcend mundane space and the general, geometrical, Euclidean space which is homogeneous in its three dimensions; again, and most important of all, *ánthrōpos* as creature, precisely on the basis of its bodily possibilities, is able to transcend the here and now. Once this is accounted for, so too is the condition for the possibility of scientific theory, e.g., Copernicus' heliocentric stance ([18], especially pp. 280–283); again, the condition of the possibility of temporal synthesis makes musical experience available to us; furthermore, all creative possibilities are grounded in the conditions of the lived-body, and we should even expect to be able to articulate the conditions of the very possibility of philosophizing itself. And if that is not enough, we may return to the 'upright posture' and determine how we constitute through it the very conditions of the possibility of moral action and even moral life!

With his unbending allegiance to the rôle of man's lived-body Straus had no need of Kant's epistemological rigour, since it was never inviting or attractive. He sought a new access to or source [*Ursprung*] of the origins of the human world which we can come to know, but he did not appeal to constituting acts of a transcendental subject that is life-less. He began again, deriving (what I prefer to call) 'categories of corporeity.' (I hesitate to use terms like 'embodiment,' ' embodied spirit,' 'incarnated consciousness,' or 'conscious incarnateness' since they all have a Cartesian ring which Straus has long ago refuted and abandoned.) Since, for Straus, man as experiencing being is creature, he has attempted, with success, to reinstate man in nature and once and for all time to 'exorcise the Cartesian ghost,' as Professor Grene has so aptly phrased it ([1], p. 195).

Since I do not intend to simply rehearse the many glories of Straus' work, I should, at this point, like to point out that what Straus seems to intend is that the relation I-Allon is precisely constituted by lived-bodily categories and not, as we have said, transcendental subjectivity or 'mind.' Thus, in effect, (though he never says it explicitly) he replaces Husserl's relation of 'intentionality' with another special relation he calls 'I-Allon.' In *Psychiatry and Philosophy* he says that the relation to the Allon 'is neither a relationship of two bodies nor of two souls; the familiar categories of cause and effect do not apply. The relation to the Allon is not calculable, nor is it statable in equations. Its distinguishing feature is the inequality and not the equality of the relata' ([16], p. 32). Here, then, we have a relation *sui generis* (the same claim strict Husserlian scholars make for the relation of intentionality). But Straus is not emulating Husserl here. He simply had had to describe a special relation that is itself achieved or constituted precisely on the condition of the lived-bodily life of man, having been physiognomically structured as he is in accordance with his own evolutionary past. Thus the I-Allon is the organized totality of objects within which *ánthrōpos* moves, that unique relation in which man becomes engaged through his motility and, eventually, his upright stance.

III. STRAUS AS PHILOSOPHER

More than a cursory glance at Straus' writings is required in order to discover the innovative reappearance of at least three traditional philosophical problems, which I should like to mention but not take up in detail. First, there is the problem of the unified self or the problem of personal identity. Straus describes the way we constitute our continuity in time although we sleep between awakened hours; after awakening [14] we realize that we have been sleeping, thus maintaining continuity through time (as well, usually, as orientation in space). How, he asks, are we justified in taking ourselves as the same person through the sleeping and wakening phases? Secondly, he raises the problem of intersubjectivity (or other minds); How is it that, notwithstanding the fact that sensory experience is fragmented, we come to apprehend a unified world of objects within the all-embracing totality of the *Allon* in which we find ourselves acting, as partners, with other? Finally, he poses the problem of the external world. We find we accept the world as permanent [*beständig*], yet we are usually told that stimuli are continuously changing. We are each rooted in our own personal and private condition, but it is equally taken for granted that we share in a public and common world in which we may, for example, witness the same tennis match.

Straus' formulations of these traditional philosophical problems (personal identity, other minds, and the external world) are stated somewhat tongue-in-cheek. For in his view these problems would never have been pursued were it not for the errors of the eminent philosophers since René Descartes. And Kant is no exception. In *Psychiatry and Philosophy* he refers to Kant's first *Critique* and reminds us that Kant 'proposes to start from the situation of man. But nowhere,' he immediately observes, 'does that great work consider us humans as live-bodies, mobile beings' ([16], p. 41). Straus correctly displays Kant's tactic, observing that the latter considers the sensory sphere as an organization which transmits a maniforld of world-free sensory data to higher 'authorities' for synthesis. Kant's criticism, Straus continues, does not consider our mode of knowing with regard to man but reconstructs it with regard to mathematical physics. Although Kant treats of space and time, he makes no mention of gravity and getting up from the ground which as *terra firma* supports us all.

Straus is, however, even a more stubborn empiricist than was Kant since he has always believed that the starting point of knowledge, or even familiar daily experience, is always the finite, concrete, particular individual. Yet Straus is no empiricist in the sense of the major British personages. The reader of a map, according to Straus, can read it not because he can associate certain data with other data but precisely because he is able to determine what is to his right and to his left once having oriented himself with respect to the map and with respect to the *terra firma* under foot. The reader is able to make the correct determination with respect to himself, the map, and the world precisely because of his own physiognomic constitution which, first of all, constitutes for him his particular 'here.' Only then can he make intelligent use of the map. But enough of spatial orientation, lived-space, and the rest.

I wish to conclude with the problem of the condition of the possibility of moral acts and moral life, to which I alluded some time ago.

IV. MAN: UPRIGHT AND UPSTANDING

For Kant, the moral world and the world of objective empirical existence are sharply separated. For Straus, nothing is further from the truth. Straus' view of uprightness is itself, retaining Kant's phraseology, the very condition of the possibility of moral actions, themselves intimately related to the 'subjective ground of distinction' of our lived-bodily lives, thus making our ability to take *a stand* possible. Straus at least supplies the missing clue to the constitution of the possibility of moral action in focusing on the lived-body in all its subtlety. He does not presume to have obtained the complete answer to this most complex problem. Permit me, then, to close with an apparent irrelevancy.

In my initial remarks I made reference to 'The Upright Posture' and the fact that it opened the way to the understanding of the constituting of the all-embracing *Allon* as well as offered an account of the condition of the possibility of the most significant human achievements – medicine, music, the dance, philosophy, science, and morality. For almost ten years one paragraph from that paper has remained cloaked in mystery and haunts the philosophical intellect, in spite of its repetition by other commentators. 'The term "to be upright" has two connotations,' Straus remarked:

> to rise, to get up, and to stand on one's own feet; and the moral implication, not to stoop to anything, to be honest and just, to be true to friends in danger, to stand by one's convictions, and to act accordingly, even at the risk of one's life. We praise an upright man; we admire someone who stands up for his ideas of rectitude [19], p. 137).

Then comes the most startling sentence: 'There are good reasons to assume that the term "upright" in its moral connotation is more than a mere allegory.'

What implications does this remark have for moral philosophy? Reflecting on this suggestion, I had occasion to formulate it when, in 1966, I spoke with William Frankena, professor of moral philosophy at the University of Michigan. I shall always remember our exchange of remarks: We were standing in the library stacks, entertaining a discussion on the current status of moral philosophy – 'Have you even considered,' I asked, 'the relation that may obtain between man's upright posture, his rectitude, and the condition of the possibility of his moral life?' I was not given a chance to qualify or rephrase my question, so absurd did it seem to Professor Frankena. He stood motionless, took as erect a standing posture as he was capable of, and said: 'Do you mean that if I do this [standing straight] I will become more moral?' 'Not quite,' I responded . . . .

For those who take lightly the gulf that separates Anglo-American from Continental philosophy, the notion that the English Channel could serve as the German Ocean was now made transparent to me. For, notwithstanding those brief moments when a *rapprochement* between these two approaches to philosophical problems seems to have been successful, one soon finds that the meaning of medical philosophical anthropology as Straus envisioned it is still quite closed to Anglo-American thinkers, in spite of the fact that Kant, well-known to all of them, published his own *Anthropologie in pragmatischer Hinsicht* in 1798 ([5], see 20).

The point of this anecdote is simply that the physiognomic characteristics which *ánthrōpos* displays have not yet been fully explored and explicated as have (say) Aristotle's and Kant's categories as reviewed by 20th century philosophers. In my view it is Straus who clearly signposted the correct path which transcendental philosophy since Kant has left untrammeled. Like Husserl, Straus is properly viewed as a radically beginning philosopher [*radikal anfangende Philosophen*] ([3], p. 48). Although all authentic philosophy must be radical, thanks to Straus we do not have to begin again *ab ovo*.

V. PSYCHIATRY AND PSYCHOPATHOLOGY

Considering the fact that Straus has for many years been a practicing psychiatrist, it is surprising how few of his publications, especially those of larger scope, deal with pathology, and how many of them are devoted to normal psychology. From this one should not conclude that Straus has deserted psychopathology. That he has not can best be seen from the 'Clinical Studies' of his selected papers in *Phenomenological Psychology*. But it does mean that in Straus' perspective little progress can be made in pathology without a fuller and broader understanding of the norm. In fact, pathology is to be developed against the background of the range of normalcy of human experience, and all pathological phenomena are to be interpreted as breakdowns in the normal relationships between the 'I' and the 'Allon.' Straus does not actually offer a complete survey of such disturbances, but he supplies enough examples – from depersonalization to schizophrenia – to show how such an understanding can work. His essays on 'The Phenomenology of Hallucinations,' 'Disorders of Personal Time in Depressive States,' 'The Pathology of Compulsion,' and 'Pseudoreversibility of Catatonic Stupor' are his own major illustrations ([19], Part III, pp. 255–339).

Straus tries to understand these pathological conditions by showing how in such situations the 'axioms of everyday experience,' which we normally take for granted, are in various ways undermined and abandoned. Thus, in the case of auditory hallucinations, the way in which, ordinarily, voices and speakers are conjoined is radically broken up: The schizophrenic hears only voices, no longer persons. Similar disruptions can occur in the tactile and even in the visual sphere: Phenomenology allows us to understand these phenomena as deformations of normal modalities.

Such examples can do no more than indicate how pathological phenomena can become phenomenologically accessible as modifications of the normal modalities of our experience. Phenomenological analysis can, perhaps, guide us to a fuller insight into what is going on in the patient's deranged experience.

In *Psychiatry and Philosophy* Straus says:

> If phenomenology is not the ultimate instrument of psychiatric theory, it is at least a powerful reminder that without philosophy psychiatry cannot make a lasting claim to knowledge. Conversely, to the extent that its cardinal insight is valid, phenomenological philosophy is a precious clue to the nature of consciousness in its normal as well as pathological modalities. ([16], Preface, p. ix).

Straus demonstrates this tactic by beginning with the problem of determining for psychiatry what is normal and abnormal. Abnormality, finding its primary expression in a breakdown, a catastrophe, leads back and makes reference to a common world, the visible world of the *Allon* [the Other]. This in turn requires an analysis of the basic, primary, animal situation [*Ursituation*]. In light of his answer, Straus then interprets psychic abnormality as a disturbance of this primary situation. This primary situation requires an analysis of the categories of man's lived-bodily being, not his so-called 'mental-life.'

VI. CONCLUSION

It was Ludwig Binswanger who, several years after the appearance of *Being and Time* (1927), tried to transform Heidegger's Analytic of *Dasein* into *Daseinsanalysen*. In his case studies, Binswanger investigated *das Dasein* of individuals, to whom he attached names like Ellen West, Lola Voss, and others. He studies their particular 'throwness' and their particular 'projects' – in short, their particular modes of being-in-the-world. Binswanger's existential analysis – whether a legitimate application of Heidegger's analytic or not – is by no means a specific psychiatric method. Existential analysis is not primarily interested in the clinical distinction of abnormal and normal behavior. In a paper, 'The Human Being in Psychiatry,' published in 1956, Binswanger claims that the psychiatrist cannot be satisfied with looking at a patient as an object, be it a person, a character, an organism, or a brain, labelled with a certain clinical diagnosis; the psychiatrist has to approach the patient as a fellowman in existential communication.

This paper leaves many questions unanswered – it may be more appropriate to say it leaves many questions unasked. 'The information on a "jacket" cannot replace the reading of those books which are demanding without being soothing' ([15], p. 142), Straus once remarked. This may be one of the reasons why in our day, although we cross the ocean in six hours, some ideas have needed over thirty years to make the same passage. I hope that the interest in Straus' work flaring up so suddenly in some quarters will not fade with equal celerity.

It is enough, then, for one paper to end here by restating not only Straus' maxim that 'human physique reveals human nature' and for that matter all the rest of the world but to repeat the words of Merleau-Ponty uttered some seven years later, in 1952, following Straus' lead – 'Rien d'humain n'est tout à fait incorporel.' Nothing human is entirely incorporeal . . . .[7]

University of Connecticut Health Center,
Farmington, Connecticut

NOTES

[1] Originally presented as 'Lived-Bodily Categories: The Constitution of I-Allon' at the Sixty-Sixth Annual Meeting of the Southern Society for Philosophy and Psychology, Tampa, Florida on April 12, 1974 at a Colloquy in Honor of Erwin Straus. Later it was revised as 'The Psychiatrist as Philosopher' and read before the members of the Department of Psychiatry, State University of New York, Upstate Medical Center, Syracuse, New York on October 31, 1974.
[2] These five conferences have been published by Duquesne University Press, Philadelphia, Penn.: I (1964), II (1967), III and IV (1970), and V (1974).
[3] See this volume, pp. 157–167. A set of Straus' writing is currently maintained in the Phenomenological Research Center of the Library of Sir Wilfrid Laurier University, Waterloo, Ontario, Canada. An earlier, less accurate bibliography of Straus' can be found in *Conditio Humana,* a *Festschrift* in Honor of E. Straus' 75th Birthday, 1966, W. von Baeyer and R. M. Griffith (eds.), Springer Verlag, Heidelberg, pp. 334–337.
[4] Personal conversation in Lexington, Kentucky, March 17, 1973. See Cairns, D.: 1973, *Guide for Translating Husserl,* Martinus Nijhoff, Hague, p. 23, Also see Husserl, E.: 1970, *Crisis of European Sciences and Transcendental Phenomenology,* D. Carr (trans.), Northwestern University Press, Evanston, I11., Part II, Sections 9h and 10; Part III, Sections 33–34, Appendix VI, p. 358.
[5] Personal conversation with Straus. See Spicker, S. F.: May 1970, 'Shadworth Hodgson's Reduction as an Anticipation of Husserl's Phenomenological Psychology', *J. Brit. Soc. Phenomenol.* 2(2), 57–73.
[6] Merleau-Ponty, M.: 1964, 'Eye and Mind', *Primacy of Perception,* J. M. Edie (ed.), Northwestern University Press, Evanston, I11., p. 162. Merleau-Ponty adopts this remark from Valéry. See Merleau-Ponty, M.: Jan. 1961, 'L'Oeil et l'esprit', *Art in France,* 1(1).
[7] Merleau-Ponty, M.: 1968, *Résumé du Cours: Collège de France, 1952–1960*, Gallimard, Paris, p. 178. For an excellent analysis of the concept of the lived body [corps propre] see Merleau-Ponty, M.: 1968, *L'Union de L'ame et du Corps chez Malbranche, Biran et Bergson: notes au cours de Maurice Merleau-Ponty, a L'école Normale Supérieure, 1947–1948,* recueilles et rédigées Jean Deprun, Librairie Philosophique J Vrin, Paris.

BIBLIOGRAPHY

1. Grene, M.: 1967, 'Erwin W. Straus', *Approaches to a Philosophical Biology,* Basic Books, Inc., New York, pp. 183–218; an earlier version appeared in *The Review of Metaphysics* **21**, 94–123, 1967.
2. Husserl, E.: 1929, 'Phenomenology', *Encyclopaedia Britannica,* London, **17**, 699–702; reprinted in R. Chisholm (ed.), *Realism and Background of Phenomenology,* Free Press of Glencoe, New York, 1960.
3. Husserl, E.: 1929, *Cartesianische Meditationen und Pariser Vorträge,* reprinted in S. Strasser (ed.), *Husserliana,* Vol. I. Martinus Nijhoff, Hague, 1950; reprinted as *Cartesian Meditations,* D. Cairns (trans.), Martinus Nijhoff, Hague, 1960.
4. Kant, I.: 1958, *Critique of Pure Reason,* N.K. Smith (trans.), Macmillan Co., London.
5. Kant, I.: 1964, *Anthropologie du point de vue pragmatique,* M. Foucault (trans.), Librairie Philosophique J Vrin, Paris.
6. Merleau-Ponty, M.: 1945, *Phénoménologie de la perception,* Librairie Gallimard, Paris; reprinted as *Phenomenology of Perception,* C. Smith (trans.), Humanities Press, New York, 1962.

7. Needleman, J.: 1967, 'Ludwig Binswanger', in P. Edwards (ed.), *Encyclopedia of Philosophy*, Macmillan Co. and Free Press, New York, 1, 309–310.
8. Spiegelberg, J.: 1972, 'Erwin W. Straus (b. 1891): Phenomenological Rehabilitation of Man's Senses', *Phenomenology in Psychology and Psychiatry: A Historical Introduction,* Northwestern University Press, Evanston, Ill., pp. 261–279.
9. Straus, E. W.: 1919, see 'Bibliography', No. 1, in this volume.
10. Straus, E. W.: 1935, see 'Bibliography', No. 22, in this volume.
11. Straus, E. W.: 1949, see 'Bibliography', No. 36, in this volume.
12. Straus, E. W.: 1951, see 'Bibliography', No. 41, in this volume.
13. Straus, E. W.: 1956, see 'Bibliography', No. 47, in this volume.
14. Straus, E. W.: 1960, see 'Bibliography', No. 59, in this volume.
15. Straus, E. W.: 1963, see 'Bibliography', No. 66, in this volume.
16. Straus, E. W.: 1963, see 'Bibliography', No. 68, in this volume.
17. Straus, E. W.: 1965, see 'Bibliography', Nos. 71 and 76, in this volume.
18. Straus, E. W.: 1966, *Phenomenological Psychology,* Basic Books, Inc., New York; reprinted as *Psicologia fenomenologica,* Editorial Paidos, Buenos Aires, Biblioteca Psicologias del Seglo XX, Vol. 25, 1971.
19. von Baeyer, W. and Griffith, R. M., (eds.) 1966, *Conditio Humana: Dr. Straus on his 75th Birthday*, Springer-Verlag, Berlin–Heidelberg–New York.

EDITORS' NOTE

The late Professor Dr. med. Erwin W. Straus consented to permit the compilation of his works, both original and in photocopy, for presentation on May 6, 1973 to the Phenomenological Research Center of the University of Waterloo, Ontario, Canada, which is now re-established at the Library of the Wilfrid Laurier University, Ontario, Canada.

The 'Bibliography' includes eighty-five entries. In addition, Straus published two volumes of collected papers: (1) *Psychologie der menschlichen Welt* (Berlin, Göttingen, Heidelberg: Springer-Verlag, 1960), hereafter *PmW* and (2) *Phenomenological Psychology* (New York: Basic Books, Inc., 1966), hereafter *PP*; the Spanish translation, *Psicologia fenomenologica* (Buenos Aires: Editorial Paidos, Biblioteca Psicologias del Seglo XX, Vol. 25, 1971), is also in print. These volumes are not registered within the numbered series since each paper in the two collections has its own entry number.

As the founder of the 'Lexington [Kentucky] Conferences on Phenomenology: Pure and Applied,' Straus edited the proceedings of these Five Conferences, which were eventually published by Duquesne University Press. At each conference Dr. Straus delivered his own paper, and they are noted as entries 69 (Volume I, 1974), 70 (Volume II, 1967), 79 (Volume III, 1970), 80 (Volume IV, 1970), and 83 (Volume V, 1974). The year of each Conference and its publication do not jibe, since publication was often delayed.

H. T. Engelhardt, Jr. and S. F. Spicker (eds.), Mental Health: Philosophical Perspectives, 157–167.

STUART F. SPICKER

BIBLIOGRAPHY OF THE WORKS OF ERWIN W. STRAUS

1. (1919) Straus, E. W.: 'Zur Pathogenese des chronischen Morphinismus', *Monatsschrift für Psychiatrie und Neurologie* **46**, 1–20. Inaugural Dissertation zur Erlangung der medizinschen Doktorwürde an der Friedrich-Wilhelms-Universität zu Berlin.
2. (1922) Straus, E. W.: 'Anthroposophie und Naturwissenschaft', *Klinische Wochenschrift* **1**(19), 958–960, May 6.
3. (1924) Straus, E. W. and Wohlwill, F.: 'Der Hitzschlag', *Spezielle Pathologie und Therapie innerer Krankheiten* **2**, 445–454.
4. (1924) Straus, E. W. and Wohlwill, F.: 'Nichteitrige Entzündungen des Centralnervensystems', *Spezielle Pathologie und Therapie innerer Krankheiten* **2**, 455–464.
5. (1924) Straus, E. W.: 'Zur Logik und Psychologie des Okkultismus', *Klinische Wochenschrift* **19**, 843–846.
6. (1925) Straus, E. W.: 'Wesen und Vorgang der Suggestion', *Monatsschrift für Psychiatrie und Neurologie*, Heft 28; *PmW*, pp. 17–70.
7. (1925) Straus, E. W. and Guttmann, E.: 'Die Nosologische Stellung der Akroparästhesien', *Klinische Wochenschrift* **44**, 1–6; reprinted in *Deutsche Zeitscrift für Nervenheilkunde* **88**, 247–253.
8. (1926) Straus, E. W.: 'Das Problem der Individualität', *Die Biologie der Person: Ein Handbuch der allgemeinen und speziellen Konstitutionslehre* **1**, 25–134.
9. (1926) Straus, E. W.: *Atlas der Elektrodiagnostik*, Stilke Verlag, Berlin.
10. (1927) Straus, E. W.: 'Über Suggestion und Suggestibilität', *Schweizer Archiv für Neurologie und Psychiatrie* **20**, 23–43.
11. (1927) Straus, E. W.: 'Dem Andenken an Richard Cassirer', *Zeitschrift für die gesamte Neurologie und Psychiatrie* **108**(5), 813–818.
12. (1927) Straus, E. W.: 'Untersuchungen über die postchoreatischen Motilitätsstörungen insbesondere die Beziehungen der Chorea minor zum Tic', *Monatsschrift für Psychiatrie und Neurologie* **66**(5/6); *PmW*, pp. 71–125.
13. (1928) Straus, E. W.: 'Das Zeiterlebnis in der endogenen Depression und in der psychopathischen Verstimmung', *Monatsschrift für Psychiatrie und Neurologie* **68**, 640–656; *PmW*, pp. 126–140.

14. (1929) Straus, E. W.: 'Über die organische Natur der Tics und der Koprolalie', *Zentralblatt für die gesamte Neurologie und Psychiatrie* **47**(11/12), 698–699.
15. (1930) Straus, E. W.: *Geschehnis und Erlebnis: Zugleich eine historiologische Deutung des psychischen Traumas und der Renten-Neurose*, Julius Springer Verlag, Berlin.
16. (1930) Straus, E. W.: 'Zwang und Raum', *Bericht über den V. allgemeinen ärztlichen Kongress für Psychotherapie in Baden-Baden*, April 26–29, p. 1.
17. (1930) Straus, E. W.: 'Die Formen des Räumlichen', *Der Nervenarzt* **3**(11), 633–656; *PmW*, pp. 141–178. The English translation, 'The Forms of Spatiality', in *PP*, pp. 3–37.
18. (1931) Straus, E. W.: 'Bemerkungen zu dem Internationalen Neurologenkongress in Bern 1931', *Der Nervenarzt* **4**(11), 661–663.
19. (1931) Straus, E. W.: 'Zur Psychologie und Psychopathologie der Sentimentalität', *Zentralblatt für die gesamte Neurologie und Psychiatrie* **62**(5/6), 399–400.
20. (1932) Straus, E. W.: 'Schlussbemerkungen zu H. Prinzhorns "Berichtigung" ', *Der Nervenarzt* **5**(3), 145–148.
21. (1933) Straus, E. W.: 'Die Scham als historiologisches Problem', *Schweizer Archiv für Neurologie und Psychiatrie* **31**(2), 1–5; *PmW*, pp. 179–186. The English translation, 'Shame as a Historiological Problem', in *PP*, pp. 127–224.
22. (1935) Straus, E. W.: *Vom Sinn der Sinne*, Julius Springer Verlag, Berlin.
(1956) Straus, E. W.: *Vom Sinn der Sinne*, 2nd ed. revised and enlarged, Julius Springer Verlag, Berlin.
(1963) Straus, E. W.: *The Primary World of Senses: A Vindication of Sensory Experience*, J. Needleman (trans.), Free Press of Glencoe, New York.
23. (1935–1936) Straus, E. W.: 'Le Mouvement Vécu', *Recherches Philosophiques* **5**, 112–138. The English translation, 'Lived Movement', in *PP*, pp. 38–58.
24. (1936) Straus, E. W.: 'Die Paroxysmale Lähmung', *Handbuch der Neurologie* **16**, 1023–1045, Verlag von Julius Springer, Berlin.
25. (1937) Straus, E. W.: 'Descartes' Bedeutung für die moderne Psychologie', *Travaux du IXe Congrès International de Philosophie*, Congrès Descartes, Paris, **3**(1–6), 52–59, Hermann et cie Editeurs, Paris. The English translation, 'Descartes's Significance for Modern Psychology', in *PP*, pp. 188–194.

26. (1938) Straus, E. W.: 'Ein Beitrag zur Pathologie der Zwangserscheinungen', *Monatsschrift für Psychiatrie und Neurologie* **98**, 63–101; *PmW*, pp. 187–223. The English translation, 'The Pathology of Compulsion', in *PP*, pp. 296–329.
27. (1940) Straus, E. W.: 'Psychological and Clinical Aspects of Space', *Journal of Nervous and Mental Diseases* **91**(5), 648–649. An abstract of a paper delivered to the Boston Society of Psychiatry and Neurology (K. Goldstein) in 1939 and printed as part of the Society's Proceedings.
28. (1940) Straus, E. W.: 'The Dead Letter and the Living Word', prepared at Black Mountain College, N.C. (December), 5 p., unpublished.
29. (1941) Straus, E. W.: 'Education in a time of Crisis', *Black Mountain College Bulletin*, No. 7. This paper is a revision of an address originally delivered on May 5, 1940, during the College's First Annual Visitor's Week and a few days before the invasion of Holland and Belgium.
30. (1941) Straus, E. W.: 'Psychology of Phobias', *Journal of the Elisha Mitchell Scientific Society* **57**(2), 196–197.
31. (1942) Straus, E. W.: 'Depersonalization: Its Significance for Psychology and Psychopathology', *Journal of the Elisha Mitchell Scientific Society* **58**(2), 124.
32. (1943) Straus, E. W.: 'Some Imponderables Influencing Morale', *Journal of the Elisha Mitchell Scientific Society* **59**(2), 112.
33. (1946) Straus, E. W.: 'The Life and Work of Karl Wilmanns', *The American Journal of Psychiatry* **102**(5), 688–691.
34. (1947) Straus, E. W.: 'Disorders of Personal Time in Depressive States', *Southern Medical Journal* **40**(3), 254–259; reprinted in *PP*, pp. 290–295.
35. (1948) Straus, E. W.: *On Obsession: A Clinical and Methodological Study*, Nervous and Mental Disease Monographs, No. 73, Coolidge Foundation, New York, x + 92 pp.
36. (1949) Straus, E. W.: 'Die aufrechte Haltung, Eine anthropologische Studie', *Monatsschrift für Psychiatrie und Neurologie*, **117**(4/5/6), *Festschrift* for Professor Karl Bonhoeffer on his 80th birthday; *PmW*, pp. 224–235. The English translation, 'The Upright Posture', in *PP*, pp. 137–165. See *Psychiatric Quarterly* **26**, 529–561, 1952. The German version is also reprinted in *Jenseits von Resignation und Illusion*, Moritz Diesterweg, Frankfurt a/Main, 1971. This volume commemorates the 450th anniversary of the founding of Dr. Straus' Gymnasium in Frankfurt a/Main, Germany. The English reappeared in R. Zaner and D. Ihde (eds.), *Phenomenology and Existentialism*, Capricorn Books, G. P. Putnam's Sons, New York, 1973, pp. 232–259.

37. (1949) Straus, E. W.: 'Die Ästhesiologie und ihre Bedeutung für das Verständnis der Halluzinationen', *Zeitschrift für Psychiatrie und Neurologie*, Vol. 182; *PmW*, pp. 236–269. The English translation, 'Aesthesiology and Hallucinations', in R. May (ed.), *Existence: A New Dimension in Psychiatry and Psychology*, Basic Books, New York, 1958, pp. 139–169.
38. (1950) Straus, E. W.: 'Die Entwicklung der amerikanischen Psychiatrie zwischen den Weltkriegen', *Archiv für Psychiatrie und Nervenkrankheiten* **184**, 133–150. *PmW*, pp. 270–288.
39. (1951) Straus, E. W.: 'Rheoscopic Studies of Expression', *American Journal of Psychiatry* **108**(6), 439–443; reprinted in English in *PmW*, pp. 289–297 and *PP*, pp. 225–233.
40. (1951) Straus, E. W.: 'The Autonomy of Questioning', *Journal of Nervous and Mental Disorders* **113**, 67–74.
41. (1951) Straus, E. W.: 'Ludwig Binswanger zum 70. Geburtstag', *Der Nervenarzt* **22**(7), 269–270.
42. (1952) Straus, E. W.: 'The Sigh: An Introduction to a Theory of Expression', *Tijdschrift voor Philosophie* **14e**(4), 1–22; reprinted in *PmW*, pp. 298–315 and *PP*, pp. 234–251. The German translation, 'Der Seufzer: Einführung in eine Lehre vom Ausdruck', *Jahrbuch für Psychologie und Psychotherapie* **2**, 113–128, 1954.
43. (1953) Straus, E. W.: 'Der Mensch als ein fragendes Wesen', *Jahrbuch für Psychologie und Psychotherapie* **1**, 139–153; *PmW*, pp. 316–334. The English translation, 'Man: A Questioning Being', in *PP*, pp. 166–187; reprinted from *Tijdschrift voor Philosophie* **17e**(1), 3–29, 1955.
44. (1955) Straus, E. W., Ferguson-Rayport, S. M., and Griffith, R. M.: 'The Psychiatric Significance of Tattoos', *Psychiatric Quarterly* **29**, 112–131.
45. (1955) Straus, E. W. and Griffith, R. M.: 'Pseudoreversibility of Catatonic Stupor', *American Journal of Psychiatry* **3**(9), 680–685; reprinted in *PmW*, pp. 335–346 and *PP*, pp. 330–339.
46. (1956) Straus, E. W., Jokl, E., and Kessler, H. H.: 'Neuromuscular Performance and Sensory Receptivity in a Triple Congenital Amputee', *Journal of the American Medical Association* **161**, 439–440.
47. (1956) Straus, E. W.: 'Some Remarks About Awakeness', *Tijdschrift voor Philosophie* **18e**(3); reprinted in *PmW*, pp. 347–363 and *PP*, 101–117.

48. (1956) Straus, E. W.: 'On the Form and Structure of Man's Inner Freedom', *Kentucky Law Journal* **45**(2), 255–269; reprinted in *PmW*, pp. 364–376.
49. (1957) Straus, E. W.: 'Der Archimedische Punkt', Tirage à part de *Recontre/Encounter/Begegnung*. Contributions à une Psychologie humaine dédiées au Professeur F. J. J. Buytendijk, Uitgeverij het Spectrum, Utrecht&Antwerpen; reprinted in *PmW*, pp. 377–397.
50. (1958) Straus, E. W.: 'Formen and Formeln', *Psychiatrie und Gesellschaft*, Hans Huber Verlag, Bern/Stuttgart; reprinted in *PmW*, pp. 398–408.
51. (1958) Straus, E. W.: 'The Brain and Human Behavior: Comments and Discussion', *Research Publication, Association for Research in Nervous and Mental Disease*, Vol. XXXVI, Williams and Wilkins, Baltimore, pp. 140–141, 271–279, 464–465.
52. (1958) Straus, E. W.: 'Objektivität', *Jahrbuch für Psychologie und Psychotherapie* **6**(1/3); reprinted in *PmW*, pp. 409–426. The English translation, 'Objectivity', in *PP*, pp. 118–133.
53. (1958) Straus, E. W.: 'Viktor E. Freiherr von Gebsattel 75 Jahre', *Der Nervenarzt* **29**(5), 233–234 (erroneously signed 'Jürg. Zutt').
54. (1959) Straus, E. W.: 'Report: The Fourth International Congress of Psychotherapy, Barcelona, Spain, September 1–7, 1958', *Psychosomatic Medicine*, new series, **21**(2), 158–164.
55. (1959) Straus, E. W.: 'Human Action: Response or Project', *Confinia Psychiatrica* **2**(3–4), 148–171; reprinted in *PP*, pp. 195–216.
56. (1959) Straus, E. W.: 'Fenomenologia Del Recuerdo Y Amnesia Infantil', *Psiquitría y Psicología Médica*, Año VII, Tomo IV, No. 2, Revista Trimestral, Barcelona, pp. 155–159. The expanded English version, 'Phenomenology of Remembering and Infantile Amnesia', *Acta Psychotherapeutica et Psychosomatica* **8**(5), 334–351, 1960; reprinted as 'Remembering and Infantile Amnesia', in *PP*, pp. 59–74.
57. (1959) Straus, E. W.: 'Victor Emil Freiherr von Gebsattel zum 75. Geburtstag', *Jahrbuch für Psychologie und Psychotherapie* **6**(4), 303–306.
58. (1959) Straus, E. W. and Lyons, J.: 'The Effects of Barbiturates in Catatonic Stupor', presented June 10, 1958 at the *Third Annual Research Conference on Chemotherapy Studies in Psychiatry*, VA Hospital, Downey, Illinois, pp. 79–84.
59. (1960) Straus, E. W.: 'The Existential Approach to Psychiatry', read as part of a Symposium on 'Man's Inner World, The Resources of Religion

and Depth Psychology', February 19, 1960, Unitarian Symposium No. 4, Cincinnati, Ohio; Published in J. H. Smith (ed.), *Psychiatry and the Humanities*, Vol. I, Yale University Press, New Haven and London, 1976, pp. 127–143.

60. (1960) Straus, E. W.: 'Über Gedächtnisspuren', *Der Nervenarzt* **31**(1), 1–12. The English translation, 'On Memory Traces', *Tijdschrift voor Philosophie* **24e**(1), 1–32, 1962; reprinted in *PP*, pp. 75–100.
61. (1961) Straus, E. W.: 'Norm and Pathology of I-World Relations', *Diseases of the Nervous System, Monograph Supplement* 22(4), 1–12; reprinted in *PP*, pp. 255–276.
62. (1961) Straus, E. W.: 'Diskussionsbemerkungen zu vorstehenden Beiträgen von W. von Baeyer, P. Matussek und W. Jacob', *Der Nervenarzt* **32**(12), 551–552.
63. (1961) Straus, E. W.: 'Signals, Signs, and Symbols', *Psychiatric Research Reports*, No. 14, pp. 1–14, December.
64. (1962) Strasu, E. W.: 'Phenomenology of Hallucinations', in L. J. West (ed.), in *Hallucinations*, Grune and Stratton, New York, pp. 220–232; reprinted in *PP*, pp. 277–289.
65. (1963) Straus, E. W.: 'Die Verwechslung von Reiz und Objekt, Ihr Grund und ihre Folgen', *Anthropologische und naturwissenschaftliche Grundlagen der Pharmako-Psychiatrie* **2**, 4–32.
66. (1963) Straus, E. W.: 'Psychiatrie und Philosophie', *Psychiatrie der Gegenwart* **1**(2), 926–994. Translated into English and printed in *Psychiatry and Philosophy*, No. 77.
67. (1963) Straus, E. W.: 'Über Störungen des Zeiterlebens bei seelischen Erkrankungen', *Zeit in nervenärztlicher Sicht*, herausgegeben von G. Schaltenbrand, Ferdinand Enke Verlag, Stuttgart, pp. 14–16.
68. (1963) Straus, E. W.: 'Zum Sehen Geboren, Zum Schauen Bestellt: Betrachtungen zur "Aufrechten Haltung" ', *Werden und Handeln*, herausgegebenen von Eckart Wiesenhütter zum 80. Geburtstag v. V. E. Freiherrn von Gebsattel, Hippokrates-Verlag, Stuttgart, pp. 44–73. The English translation, 'Born to See, Bound to Behold', in *Tijdschrift voor Philosophie* **273**(4), 659–688, 1965. This article is reprinted in S. F. Spicker (ed.), *The Philosophy of the Body*, Quadrangle Books, Chicago, 1970, pp. 334–361.
69. (1964) Straus, E. W.: 'Chronognosy and Chronopathy', *Phenomenology: Pure and Applied*, Duquesne University Press, Pittsburgh, Penn., pp. 142–165; also opening remarks of the Conference, pp. 3–9.
70. (1964) Straus, E. W.: 'Über Anosognosie', *Jahrbuch für Psychologie und*

Psychotherapie **11**(1), 26–42. The English translation, 'Anosognosia', in *Phenomenology of Will and Action*, Duquesne University Press, Pittsburgh, Penn., 1967, pp. 103–126; also comments of Dr. Abraham Wikler, pp. 127–131; Dr. Straus, pp. 135–137. 'Anosognosia' is a revised version of 'Über Anosognosie'.

71. (1965) Straus, E. W.: 'The Expression of Thinking', in J. M. Edie (ed.), *An Invitation to Phenomenology*, Quadrangle Books, Chicago, pp. 266–283.
72. (1965) Straus, E. W.: 'The Sense of the Senses', *The Southern Journal of Philosophy* **3**(4), 192–201; reprinted in D. Van de Vate, Jr. (ed.), *Persons, Privacy and Feeling: Essays in the Philosophy of Mind*, Memphis State University Press, Memphis, Tenn., 1970.
73. (1966) Straus, E. W.: 'Dem Andenken Ludwig Binswangers (1881–1966)', *Der Nervenarzt* **37**(12), 529–531.
74. (1967) Straus, E. W.: 'An Existential Approach to Time', *Annals of the New York Academy of Sciences* **138**(2), 759–766.
75. (1969) Straus, E. W.: ' "Discussion" of Dr. Lehmann's existentialist view of dream interpretation', in M. Kramer (ed.), *Dream Psychology and the New Biology of Dreaming*, Charles C. Thomas, Springfield, Ill., pp. 165–171.
76. (1969) Straus, E. W.: 'The Polarity of Sensory Experience – The Spectrum of Senses', *Journal for the Study of Consciousness* **2**(1), 24–35. Presented at the First International Conference on Consciousness and Creativity, the St. Vincent Medical Center, New York City, October 1967.
77. (1969) Straus, E. W.: 'Embodiment and Excarnation', *Psychological Issues* **6**(2), Monograph **22**, 217–236; 'Discussion', pp. 237–250. This article is a reworked and expanded version of 'The Expression of Thinking'. It was originally delivered at a meeting in 1965 (see No. 71).
78. (1969) Straus, E. W.: 'Psychiatry and Philosophy', in M. Natanson (ed.), *Psychiatry and Philosophy*, Springer-Verlag, New York, pp. 1–83; also the 'Preface' by Straus and Natanson, pp. v–ix. This is an English translation of No. 66.
79. (1970) Straus, E. W.: 'Phenomenology of Memory', *Pure and Applied Phenomenology*, Duquesne University Press, Pittsburgh, Penn., pp. 45–63.
80. (1970) Straus, E. W.: 'The Phantom Limb', *Phenomenology Pure and Applied: Aisthesis and Aesthetics*, Duquesne University Press, Pittsburgh, Penn., pp. 130–148.

81. (1970) Straus, E. W.: 'The Miser', in J. M. Edie, F. H. Parker and C. O. Schrag (eds.), *Patterns of the Life World*, Northwestern University Press, Evanston, pp. 157–179.
82. (1971) Straus, E. W., Aug, R. C., and Ables, B. S.: 'A Phenomenological Approach to Dyslexia', *Journal of Phenomenological Psychology* **1**(2), 255–235.
83. (1973) Straus, E. W.: 'Sound, Words, Sentences', *Phenomenology, Pure and Applied: Language and Language Disturbances*, Duquesne University Press, Pittsburgh, Penn., pp. 81–105. Delivered on April 13, 1972.
84. (1975) Straus, E. W.: 'The Monads Have Windows', *Phenomenological Perspectives: Historical and Systematic Essays in Honor of Herbert Spiegelberg*, Martinus Nijhoff, The Hague, pp. 130–150.
85. (1975) Straus, E. W.: 'L'Observateur Oublié' *Present à Henri Maldiney*, Éditions L'Age d'Homme, S. A. Lausanne, pp. 235–248.

MISCELLANY

1. (1942) Straus, E. W.: 'A Proposal for the Education of Veterans (GI Bill)', prepared at Black Mountain College, N. C., unpublished.
2. (1961) Zutt, Jürg: 'Erwin Straus – 70 Jahre', *Der Nervenarzt* **32**(10), 437–438.
3. (1966) von Baeyer, W. and Griffith, R. M. (eds.): *Conditio Humana: Dr. Straus on his 75th Birthday* (Festschrift), Springer-Verlag, Berlin–Heidelberg–New York.

PHILOSOPHICAL ANALYSES

1. (1968) Grene, M.: 'Erwin W. Straus', *Approaches to a Philosophical Biology*, Basic Books, Inc., New York, pp. 183–218.
2. (1972) Spiegelberg, H.: 'Erwin W. Straus (b. 1891): Phenomenological Rehabilitation of Man's Senses', *Phenomenology in Psychology and Psychiatry: A Historical Introduction*, Northwestern University Press, Evanston, Ill., pp. 261–279.

OBITUARY

1. (1975) Eng, E.: 'A Man of Sensitivity', *Sunday Herald-Leader*, Lexington, Kentucky, May 25; excerpted from the 'Eulogy for Erwin Straus' offered by E. Eng on May 22, 1975 in Lexington.

2. (1975) 'Doctor Who Fled Germany, E. W. Straus Dies at 83', *The Lexington Leader*, May 20, p. 18.
3. (1976) Bräutigam, W.: 'Erwin Straus, 1891–1975', *Der Nervenarzt* **47**, 1–3.
4. (1976) Spiegelberg, H.: 'Erwin Straus (sic) (1891–1975)', *Journal of the British Society for Phenomenolgy* 7(1), 69–70.

ROBERT NEVILLE

ENVIRONMENTS OF THE MIND

The relations between a person's awareness and his body have been problems for modern thought ever since Descartes pointed out that each can completely be conceived without essential reference to the other. Whatever thought may be and however wide its range, no thoughts are spatially extended; even the thoughts *of* spatially extended things are not spatially extended. Similarly the realm of extended things may be of whatever character physics discovers it to have, yet none of those extended things need be conceived to think. Descartes' own resolution of the problem broke down in two related ways. He suggested that mind and body are two distinct substances that interact through conjunction in the pineal gland. Yet if the mind were casually related only to mental elements, it did not seem possible for it to move the pineal gland; or if it could, then why not other bodily parts also? Furthermore, most people have come to believe that whatever a person is, he is one thing, not two essentially different things; therefore the mind–body problem now is one of conceiving of two somehow different aspects of a person. Put in more contemporary language, the problem is to understand the relation between physical and mental functions.

The mind–body problem is more than a conceptual problem at the present time. In the clinical treatment of behavioral disorders it is possible to alter mental factors of personality by intervening in physical processes. Psychosurgery and psychopharmacology are examples. Furthermore we are coming to understand something of the anatomical, chemical, and electrical processes involved in learning and memory, and perhaps the experiential aspects of education can be altered by physical interventions. Because we need to establish moral limits and rationales for such interventions, there is an immediate practical problem now of understanding the relation between mental and physical functions.

Standard philosophic strategies for solving the mind–body problem have been either of Descartes' sort, suggesting that there are two substances interacting, or of a reductionist sort, suggesting that mind is really a brain state or that physical objects are special kinds of minds. From a clinician's point of view, however, the problem is the interaction of two elements of the same person; the person is just as much his body as his mind.

H. T. Engelhardt, Jr. and S. F. Spicker (eds.), Mental Health: Philosophical Perspectives, 169–176.

My purpose in this introduction to Dr. Feldstein's paper is to suggest a 'non-standard' philosophic strategy by drawing out the doctrines of two philosophers whose solutions to the mind–body problem have never seemed plausible or practical, namely, Spinoza and Leibniz. The reasons they have not been taken seriously on this point are that their separate doctrines are incomplete and that they seem to offer no categories that connect with practical experience. In the 20th century, however, the philosopher Whitehead presented a theory that brilliantly combined their doctrines; and the developments of science have brought us head-on to practical problems that cannot be expressed in common sense terms but rather in terms more coordinate with Spinoza and Leibniz. The paper by Professor Feldstein deals with one such problem, the distinction of the ways in which the unconscious mediates bodily and mental functions in normal people from the ways it does in neurotics. My intention here as a preface to his paper is to present the philosophic context in which it rests.

From Spinoza we may borrow the doctrine that a thing may be understood as having a physical nature and a mental nature. This is true of everything, according to him, although of course the mental natures of most things do not have interesting thoughts. While it may seem implausible to say a stone or electron has mental functions, let me suggest that this is the most elegant and simple hypothesis in the long run. Furthermore, because of its two natures, each thing is related to other things in two ways. It is related to physical things according to spatio-temporal properties and physical causal laws; the extent and kind of physical determination is an empirical matter, and no part of physicalistic empirical science need be denied in principle. By virtue of its mental functions each thing is related potentially to other things by mental properties that express themselves in people as intending, remembering, symbolizing, and so forth; the various characters of mental functions are also empirically to be discovered. The mental faculties of human beings are probably very refined versions of the mentality of other things; intention and symbolism of things beyond the immediate body are probably limited to animals, for reasons I shall suggest below. The mental properties of electrons are likely no more different from those of people than the physical properties of electrons are from human bodies.

From Leibniz we may borrow the doctrine that a person is not a single substance but a society or ecosystem of minute changes. Some ecosystemic relations are those we describe as physical; others are mental. Whereas Leibniz hypothesized that a person is an ecosystem of infinitesimal enduring monads, I suggest, following Whitehead, that each of these 'changes' is a momentary happening, an 'occasion.' The continuity of a person, apart from moral relations, consists not in his being the same substance from one day to the next, but rather in his exhibiting more or less the same ecosystemic patterns. The patterns remain continuous or change by acceptably slow increments, even though each occasion within the person has just a momentary present existence. Ordinarily we take those patterns making up an organized physical body to determine the scope of a person; but sometimes we identify a person with his actions that extend beyond his body. There are also some gray areas where ecosystemic interactions may be said to be either part of the person *or* part of an external environment. Which relations are said to remain within a person, and which extend beyond him depends on the pragmatic interests involved in determining the person's limits.

Putting together the contributions of Spinoza and Leibniz, let me suggest that a person is a set of occasions with various ecosystemic relations among themselves and with the surrounding environments. Our immediate problem is that some of these relations are physical and some mental. The most interesting relations between human occasions, however, combine physical and mental properties, for instance, those involved in gesturing. A semantic relation or system involved physical occasions interpreted by other occasions as signs that can be subjectively enjoyed by some mind. A semantic system that involves no appropriate mentality, e.g., an unread book, is not significantly mental. An act of mental awareness takes place in a nest of environments. Some of these are physical, for instance, those constituting the brain, its past functioning, and its systematic connections with the body and the rest of physical nature. Others of those environments are semantic, involving the meanings of memories and intentions. An action is a train of occasions through both physical and semantic environments. One way of characterizing consciousness is to say that it involves a nest of physical relations with semantic connections that can be subjectively formed.

An occasion should be understood as exhibiting roles in all the ecosystems in which it plays parts. In fact, it is an 'existential obligation' of each occasion, as it were, to reconcile all the roles it has to play if it is to occur in its particular nest of environments. The possibilities of an occasion are set by the conjunction of all its environments, physical and mental. System-theory offers tools for analyzing this in detail.

Furthermore, the way an occasion plays a role in one ecosystem affects the ways it plays roles in others. To put it graphically with respect to mental and physical roles, how you think of your boss depends on whether the occasions of those thoughts take place in a drunk brain or sober one. Semantic roles are determined in part by the fact the occasions playing them are also playing physical roles in a brain with varying states; and vice versa.

The critical test for this hypothesis, typical of process philosophy, is whether sense can be made of the claim that an occasion can have both physical and mental functions. Merely to *say* it does is to name the mind–body problem, not solve it. To suggest an explication of the mental and physical aspects of an occasion permit me to make three points.

First, there is a difference between the way a thing feels its present self and the way it is felt or perceived by others, or by itself at a later time. The former is the subjective immediacy of being the occasion as it happens; the latter is the objective character the occasion has ready to be grasped by another occasion. The subjectivity of an occasion perishes when its present time is past. And the objective character does not exist until the occasion has happened. My first point is that the subjective feeling of being a brain state is the root of mental awareness; its objective character, with one important kind of exception, is its physical properties as a brain state.

Second, the subjective feeling of being an occasion has a definite set of requirements to which it must conform, namely, those structuring the process of combining previous occasions, which are its raw material, into a consistent singular, definite outcome, its own physical self. An occasion, from its subjective side, is the making of a new happening out of many things that have happened. Whitehead and a school of criticism taking its rise from his ideas have worked out the logic of this process in great detail.

In the coming-to-be of an occasion, patterns of possible harmony are abstracted from previously given occasions. If the past environing occasions are rather homogeneous, as in the center of a rock, for instance, an emergent occasion is limited to harmonizing itself according to patterns very much like those of its neighbors. In a rich environment such as the human brain, well formed by complex social experience, the variety of patterns that may be abstracted and combined is far greater. Whitehead called these possible patterns 'propositions,' both because they function within the subjective process as proposals for the occasion to use in its self-unification and because in principle they can be represented as propositional functions in symbolic logic. If you ask what contribution an occasion makes to the past conditions out of which it arises, the answer is the subjective process of abstracting out propositions, of combining them into consistent possibilities for bringing the happening to fruition, and realizing one such possibility so that something actually happens.

To speak of an occasion being 'mental' and having subjectivity need mean no more than that an occasion constructs itself according to a pattern derived from an immediate predecessor, as an occasion within a rock repeats the pattern of its environing occasions. Human mentality by contrast involves intention and symbolism because the body and the way the body perceives the environment involve some semantic relations.

My third point then is that certain occasions, such as some of those taking place in the human brain, have the capacity to abstract out of certain predecessor occasions propositions of a subjective nature. Most occasions abstract only the objective physical patterns of their predecessors, we must suppose. But certain complicated occasions enjoying the physical environment of a brain can abstract out those propositions that themselves played abstract roles in the subjective genesis of the predecessor occasions. Although the subjective immediacy of those past occasions has perished, the subjective feelings of process as embodied in their physical characters remain to be felt subjectively by certain successor occasions. As a consequence, there can be a continuity of subjective feeling from one occasion to the next, so that certain logical relations between signs can make the propositions of one occasion interpret those of others. Human mental life requires a brain, not a rock, for its physical environment because a brain allows for occasions whose propositions relate semantically, according to ecosystems of meaning. Let me attempt to make this more plausible.

Certain trends in modern linguistics suggest that the deep structures of meaning are indigenous to the human brain by virtue mainly of genetic inheritance. What can this mean? That certain physical activities of the brain are such that when they are happening the person whose activities they are is having semantically structured experience that expresses itself in some language or other. Language of course need not be verbal, only an affair of signs or meaning relations. The meaning relations between mental events are not at all the same as the physical relations between brain events, despite the attempt of computer models of mental activity to suggest the contrary. Rather, the relation between mental contents is to be understood in terms of a theory of signs, however bizarre some of the sign systems seem. As the linguists suggest, there may be layer on layer of signification and translation underneath the signs of which we are explicitly conscious, and many layers may play roles in awareness, as Dr. Feldstein shall argue. The relations between brain events are to be understood in terms of anatomical pathways, electrical excitations, and chemical reactions. It would be reductionistic to reduce either to the other.

But we can easily understand how each might require the other. It seems plausible that only those occasions whose propositional patterns are in semantic ecosystems so that they can be interpreted by other propositions have subjective feelings that can be felt by later occasions. By contrast, the propositions in occasions in a rock are merely patterns of previous occasions by which those occasions organize themselves; they do not mean one another in any kind of interpretive system. But some of the propositions in some brain occasions do stand in interpretive relations to each other, with the consequence that subjective continuity is possible. Only a physical environment as complex as a brain seems to allow for propositions of a symbolic sort; part of the complexity consists in simplifying and organizing the coordinated impulses from the rest of the body and surrounding world so that the brain occasions can pattern themselves according to what is important. A brain such as humans have would not have evolved, we may suppose, if it were not for the adaptive value of certain symbolizing functions that require such a brain organization. Neural connections between processing centers for speech and vision, for instance, would have no adaptive value were it not for the symbolic relations between sights and words.

It should be remembered that each occasion plays many roles in many ecosystems constituting its environment. It must find ways of reconciling these roles if it is to be possible at all. Some of the ecosystems are physical systems, responsive to physically understood happenings. Others are semantic ecosystems, understood in terms of interpretations. But the roles an occasion plays in one ecosystem affect how it relates to the others.

Semantically meaningful human experience should not be confused with the continuity of subjective feeling – mental life in the strict sense. Subjective continuity is a problem for any theory obliged to say that one brain state 'remembers' another, and I have addressed that problem on behalf of the theory of occasions. Meaningful human experience, for instance, gesturing or speaking out loud, involves the physical aspects of occasions – the wave of the hand or the noise – as well as the subjective interpretation. For understanding most aspects of human life it is only obfuscatory to distinguish physical and mental elements. Rather we should understand the semantic ecosystems and their environments. The chief reason to distinguish mental and physical aspects is to cope with Descartes' observation that thoughts are unextended. The point is that the human thinking of a thought must be seen in a much more complex environment.

Human beings are enduring entities with personal structures that change slowly. But what might the term 'structures' mean here? If we adhered to a theory of substances we could explain structures as the patterns according to which the substance is shaped and moves. What is it that is shaped and moves, however? The answer must be in terms of some kind of pre-structured matter, for instance, physical matter or an ineluctable mental matter; this part of the logic of substance is what led Descartes and others to speak of physical substance, with physical structures, and metal substances with mental structures. On the theory of occasions, however, it is not necessary to specify a pre-structural matter; the occasions are simply happenings whose structures are ways by which the various environments within which each occasion takes place are reconciled.

An enduring person, however, cannot be understood in terms of the structures of his or her constituent occasions alone, for the enduring continuities are some of the most important elements of personal structure. Instead, the enduring structures are repeating patterns embodied by successive nests of occasions. In an organic body the patterns do not repeat in every occasion but repeat in interrelated rhythmic ways, like the beat of the heart. Rhythmic systems are not isolated; they interact, so that muscular exertion, for instance, makes the heart beat faster, excites rudimentary feelings of fight or flight, and perhaps alters the galvanic skin response. Endurance is a function of repetition of pattern in a rhythmic way. The endurance of complex interrelated system involves highly complex interacting rhythms, some of which are constituents of others. All enduring structures, it appears, exhibit the rhythms of electromagnetic fields. Some organize electromagnetic rhythms into the rhythms of 'solid bodies;' break the solid body into elementary constituents, however, and you are back to the dance of electrons. Semantic structures are rhythms of a very refined and subtle sort. Human mentality involves the very special rhythm described above in which the subjective propositions of one occasion are reiterated in subsequent occasions.

Professor Feldstein's task is to describe certain kinds of structures, or rhythmic patterns, that are important for human life. In particular, he describes those grossly called 'bodily' and those called 'mental' and connects them with those called 'unconscious.' His special interest, however, is not these rhythms in isolation, but rather their effects upon one another. How do bodily rhythms condition mental ones? How do mental ones condition the bodily? His answer is, through the special rhythms of the unconscious.

The relation between rhythms is a peculiar one. We may describe separate structures as if they could in fact be systematically isolated from one another; we may then compare or map them onto each other to see whether they are congruent, or synchronous. Do the rhythms beat together? Do they require an intervening rhythm to make them harmonious? To answer these questions requires developing the metaphors by which we ordinarily describe harmonies into somewhat technical terms. A musical note is an audible vibratory rhythm of a certain number of cycles per second. Two notes sounding at once reinforce or inhibit each other's rhythms; two discordant notes may be brought into harmony by a third note, or two concordant notes may be thrown into discord by a third. For Professor Feldstein, the unconscious is something like that 'third note,' harmonizing various personal rhythms in a healthy person, creating discord in an unhealthy one.

Professor Feldstein uses three main sets of metaphors. One is based on 'gravity' and allows him to speak of levitation and weightiness; if one conceives a person as he does to be a hierarchy of rhythmic structures, levitation refers to certain elements of patterns lower in the hierarchy being exemplified in higher structures; weightiness is the reverse. The metaphors of refined and coarser rhythms are self-explanatory. The metaphor of luminosity (and obscurity) is at the heart of his argument; it refers to the capacity of any one rhythmic structure to reflect others so that those others can be perceived in it. A luminous structure is like a window in that it reveals something else. It is like a light in that because of its special quality, something can be seen that could not be seen without it; the unconscious reveals bodily needs, for instance, of which a person would otherwise be unaware.

The various rhythmic structures of a person can be viewed either objectively or subjectively, depending on what instruments of vantage we have. Professor Feldstein's main concern is to create a language or a subjective view. His metaphors therefore aim to evoke experiences of self at the threshold of consciousness. Our usual language distorts those experiences. The concreteness and relevance of Professor Feldstein's metaphors can be appreciated by bearing in mind that they apply to the difference between mental health and illness as he interprets that.

State University of New York at Stony Brook,
Stony Brook, N.Y.

LEONARD C. FELDSTEIN

LUMINOSITY: THE UNCONSCIOUS IN THE INTEGRATED PERSON

In this paper,[1] I propose three theses: first, located in processes midway between bodily activity and mental activity, the Unconscious mediates transformations from each sphere into the other; secondly, revealing these processes in every aspect of his behavior, the healthy person exhibits one mode of unconscious functioning, whereas the unhealthy person, whether ill of body or ill of mind, exhibits a quite different mode; finally, man's characteristic searchings are themselves conditioned by this status of the Unconscious and, in particular, by the dialectical interplay, insofar as both modes occur within him, between the wholesome mode and the unwholesome mode. Accordingly, I stress these areas of inquiry: the locus of the Unconscious, its agential character, its *modus operandi*, its internal dynamics, and its presence in human comportment. My aim is to illuminate the role of the Unconscious in the actions of integrated persons; and my chief conclusion will be that the Unconscious glows, as it were, through those actions and confers upon them a special quality that I call *luminosity*.

I. THE UNCONSCIOUS AS MEDIATOR

In his studies on the philosophy of mind, Charles Sanders Peirce wrote: 'Our whole past experience is continually in our consciousness, though most of it sunk to a great depth of dimness' ([7], 7.547). For, he continues, 'consciousness is like a bottomless lake in which ideas are suspended at different depths,' a lake 'whose waters seem transparent, yet into which we can clearly see but a little way' ([7], 7.547). Indeed, the more profoundly one plumbs these depths, the more one sinks into the abyss of un-consciousness. Beyond that, as one descends into ever more dim regions of consciousness, and passes the threshold of that which is even recallable, one in effect descends into the body itself. For, as I argue, in its grounding depths, consciousness *is* body. Sufficiently penetrated, it reveals body not only as its substratum dissociated from, though supporting, mental activity, but, in addition, as suffusing that activity and invading its every form. For pervading all mentation is not only the awareness of body, however dim and buried that awareness be, but body itself.

H. T. Engelhardt, Jr. and S. F. Spicker (eds.), Mental Health: Philosophical Perspectives, 177–189.

How can the body pervade the mind? Just as our whole past experience registers itself in consciousness, though its original vivacity is greatly diminished, so it registers itself as thus having perished yet always abiding within every region of its nethermost stratum, the Unconscious. *A fortiori*, these depositions engrave themselves even in the stratum which lies below unconsciousness, namely, the very processes that compose the body. As mind sinks into labyrinthine recesses of body, it acquires something of the same arcane character. Indeed, Freud often stresses the bizarre sinuosities of unconsciousness, its strange meanderings and grotesque formations.[2] According to my thesis, consciousness *means* body. For it *is* body as the latter mirrors itself to itself, reflecting into its own constitution the congeries of relationships it sustains to other bodies. Surely, the body which a particular awareness experiences itself as possessing is integrally enmeshed with bodies external to it, bodies some of which are likewise associated with other awarenesses. And the sphere of mental activity expresses this interwoven physical fabric of existence under the perspective of a particular body. The variegated facets of human comportment work out its diverse intentions. Consciousness is grounded and, indeed, constituted by the carnal.

It is equally true to speak of body as enveloped by consciousness. For, when it has acquired sufficient intricacy amid the growing unity of its organization, body manifests its activities in the symbols of consciousness, the images by which awareness re-presents to itself the corporeality that is primordially present to it. Inscribing upon its own processes resonances far more subtle than any that may be detected by known physiologic means, body in effect etherealizes itself and transfigures its own status from nonreflective to reflective. For when the rhythms of its processes have become orchestrated in a manner sufficiently complex, and when in their intricacy these rhythms constitute echoes which reverberate throughout the body, in a fashion that brings about ever more cohesive patterns of bodily integration, the realm of mind is born.[3] It is as though the body's rhythms have been buoyed up into mentation; it is like the lifting of the body against gravity that had, in effect, held it down as mere body.

The Metaphors

In discussing the distinction between the mental and the corporeal, I interweave three sets of metaphors: levitation and weightiness, refined and coarser, luminosity and obscurity. The coarser rhythms are those which, so to speak, press the person down into his own corporeality. There, among body's obscure and labyrinthine recesses, he dwells – a Caliban entrapped within a physique that struggles but cannot quite emerge into the luminous spheres of consciousness. But too, Caliban declares,

> Be not afeard. The isle is full of noises,
> Sounds, and sweet airs, that give delight and hurt not.
> Sometimes a thousand twangling instruments
> Will hum about mine ears, and sometimes voices,
> That, if I then had wak'd after long sleep,
> Will make me sleep again; and then, in dreaming,
> The clouds methought would open and show riches
> Ready to drop upon me, that, when I wak'd,
> I cried to dream again.
>
> Shakespeare, *The Tempest*, Act III. Scene 2, 144–152

So the most marvellous rhythms dwell within the human body, rhythms imprisoned therein until a body awakens from its slumbers and achieves the unity of selfhood. Within a field of primordial silence, the stillness of *mere* bodily activity, fragmentary experiences, each an evanescent appearing, each a passage from mere potentiality to actuality, and each anchored in specific body processes, all unfold and come to confluence as an integrated manifold. Within this complex, the coarser rhythms cling together and, as it were, congeal to form mental structures that are ineluctably unconscious.[4] Those which are more refined constitute the more evanescent though less tangible realm of consciousness.

Rhythms and Strata

In my account, I distinguish three levels of human being, each itself composed of subordinate strata: the rhythms of consciousness, the rhythms of the Unconscious, the rhythms of body. How these rhythms are intertwined is the topic of my present deliberation. With respect to the rhythms of body, I further distinguish three (physical) strata: the embeddedness of body within other bodies, the structures of body, the functions of bodily structures. And with respect to the rhythms of consciousness, I likewise distinguish three (mental) strata: simple awareness, self-consciousness, unselfconsciousness. The rhythms of the Unconscious are interposed between those two sets of strata. Exhibiting their own intricate mode of lamination, these rhythms are, as it were, the media through which mental phenomena sink into the physical and the latter rise to the condition of the mental. To elucidate the dynamics of this activity, I first consider certain analogies between the structure of body and the structure of mind.

Bodies: Connected and Independent

Human body is profoundly implicated with other bodies. The interplay of these physical events exhibits the character of physical space, time and matter. In the course of its development, influences flowing from diverse natural complexes converge upon specific loci to constitute a relatively autonomous configuration, a (human) body that in effect is striving (by a kind of immanent intentionality) to raise itself out of its own embeddedness with other bodies. Causally efficacious factors contribute their diverse impacts upon this configuration. They imprint their own form, and transmit to it a portion of their indigenous rhythms. Weaving these rhythms to a specific unity of action, every *particular* body contributes its own ornamentation. Thereby, it enables itself to stand forth and so evolve as to declare its relative independence amid its welter of dependencies. In this standing forth, structures emerge which are tangible, durable and concrete. Yet, to be woven together in such fashion that the body does not lapse into passive embeddedness, these structures must, in effect, 'functionalize' themselves. For structures must melt into a ferment of interwoven paths, paths that are not morphologically distinct but, on the contrary, fluid and in perpetually shifting relationships. The brain is an excellent example of an organ whose components are not sharply delimited. For it to function efficaciously, all parts must be mobilized in order that a single pathway be activated for the purpose of transmitting some neural influence. To a lesser extent, all bodily organs exhibit the same relative amorphousness of structure, though quite definite design of functional interdependencies.

Feelings

Once, however, man's 'glassy essence' ([7], vol. 6, p. 155) has liquefied, altogether new modalities of functional activity supervene; mind is born. Once again, to cite Peirce,

> Protoplasm, when quiescent, is, broadly speaking, solid; but when it is disturbed in an appropriate way, or sometimes even spontaneously without external disturbance, it becomes, broadly speaking, liquid . . . (And as such) . . . protoplasm feels. It not only feels but exercises all the functions of mind . . . (moreover) . . . the feeling at any point of space appears to spread and to assimilate to its own quality, though with reduced intensity, the feelings in the closely surrounding places. In this way, feeling seems directly to act upon feeling continuous with it ([7], 6.247ff).

According to this view, which I here support, living body, an organism, when sufficiently organized amid the 'liquefaction' of significant systems constituting it, is the locus of spreading feelings, feelings that emerge as expressions of the subtler resonances that are transmitted through those systems. As such, feelings, quite literally, ex-press, in the sense of pressing out, those resonances and press them into adjacent or neighboring bodies.

Feeling not merely accompanies bodily acts, and bodily interaction, but, beyond that, feeling is the realm of functions inscribed upon body as it thus acts and inter-acts. It is the echoing of body, that orchestration of rhythms which comprises body resonating to itself and thus producing systems of intraorganismic vibration.

Self-Consciousness

Consciousness of self emerges when one set of resonances crystallizes as the sense of *I-ness*. A focal region within the finely modulated resonances of awareness, this *I*-ferment is a relatively circumscribed, relatively autonomous configuration which itself endures as the invariant core of awareness even while it itself resonates to the vibrations that sweep about it. In self-consciousness, the self mirrors itself to itself; and thereby it duplicates the reflective propensities of body itself. In primary reflection, the body orients itself toward its ramifying connections with other bodies. In derivative reflection, reaching its culmination in self-consciousness, body orients itself (secondarily) toward itself in the form of schemes of symbols by which it interprets, in ever new layerings of meaning, its own feelings as they are directly and intimately associated with specific acts of relatedness.

However, by self-consciousness, I do not mean an exclusively monadic and, as it were, narcissistic construal of the processes of mind. For, all mental phenomena are intrinsically dyadic. They are shot through with resonances derived from *other* minds, resonances transmitted as persons interlock with one another; and their respective mental vibrations, in effect, lock-in to each other. For as persons actively turn toward one another, they, so to speak, direct their mental resonances toward one another; they orient their minds in each other's way. But this gripping of one person by another occurs essentially in the stage where self-consciousness has been superseded by *un*selfconsciousness. Herein is replicated on the level of mind a phenomenon that first revealed itself on the level of body – in the phase of embeddedness.

Personal Autonomy

There are two poles of intimacy of relatedness: the embeddedness of body, the embeddedness of mind.[5] Between these poles, the person achieves a measure of autonomy. He works out his destiny, passing from body structure to body function, from unconsciousness to consciousness and self-consciousness: a succession of metamorphoses and transfigurations. It is equally true to say that the body is weighted with unconscious residues as to say that mind is permeated by unconscious efflorescences. For the

Unconscious plays a dual role. On the one hand, when the Unconscious, as it were, jells, it cannot rise into awareness; it thickens into relatively rigid structures which merge with those of body itself. On the other hand, when the Unconscious liquefies, becoming less viscous, it allows for the smooth and synchronous passage of impulses from each sphere into the other. Body becomes, so to speak, luminous with the rhythms of mind; its several parts are well-coordinated.

II. THE UNCONSCIOUS IN HEALTH AND IN PATHOLOGY

Physical Pathology

In physical pathology, the diseased organ functions as though it were an autonomous system, undergoing vicissitudes in accordance with principles peculiar to it alone. Walled off from the remainder of the body, it evolves through the action of indigenous dynamisms which, on superficial inspection, are unrelated to those characterizing the healthy organism. But deeper examination reveals an inner connection between these seemingly independent factors. For wherever there is disease, the afflicted organism discloses itself to be governed by two sets of norms. The first set expresses the interplay between two relatively enduring configurations of processes in their apparent mutual autonomy; the second set expresses a more inclusive single system whose parts are more diffusely bound together. In effect, the organism is bifurcated into two compresent schemes of action. The connection between these schemes is intricate and often labyrinthine. When, however, pathology is rectified, but a single scheme presents itself, and the network of relationships characterizing the dual state vanishes, an economy of functioning prevails. Coherence in the lineaments of the manifestations of functioning replaces the inchoate state that hitherto had been operative.

Mental Pathology

When a person's psyche has been traumatized, two foci of potential consciousness supervene. One presents itself to the world as the locus of overtly intended behavior; the other presents itself as the locus of concealed tendencies – so to speak, a subterfuge with respect to the first locus. I am referring to Freud's *repressions*, which express the diseased Unconscious. In effect, processes have been dissociated from consciousness and related to a peripheral though not irrelevant 'region,' a region haunting consciousness yet functioning by its own characteristic laws. Should *psychic* pathology be rectified, the dual system is replaced by a more concretely unitary and monadic system; the linkages dissolve and a single, cohesive psychic scheme presents itself.

Luminosity

Accordingly, I am proposing several theses. First, all physical events, hence physical pathology, have their psychic correlates; and conversely. Next, if the diseased tissue in physical pathology may be regarded as functioning autonomously, so does the diseased psyche in mental pathology. As in the case of physical pathology, the psychic disabilities likewise register themselves within the rhythmic complex that is the Unconscious. In the latter case, one may speak of cure as the displacement from a psychic bifurcation into consciousness and from the Unconscious to a spontaneous unselfconsciousness. Finally, when unselfconsciousness is in perfect correspondence with this ease of comportment, one may speak of the *luminosity* of the person.

The meaning of this concept must now be examined in the light of the theory of correlationism and transformation from one psycho-physical state to another, i.e., from the diseased state to the 'normal' state. In effect, I am suggesting a concept of emotional (as well as, by implication, physical) well-being. To deal with this problem, I again advert to the nature of body and to the nature of mind. In both instances, activity is based on a principle of association.

Bodily Integration: Attachments and Detachments

The physical events constituting the body are bound together into a unity of organization. What holds its parts together is the workings of a principle according to which events show pronounced affinities for other events – i.e., when the former have previously been associated with events like the latter. For example, both replication and immunologic responses are based on the fact that certain molecular groupings on some macromolecules recognize other molecular groupings on other macromolecules.

Similar events (i.e., events belonging to the same class) have previously so imprinted themselves upon the given events as to have been woven into their actual internal constitution. In consequence, detachment from such events is felt by the given events as a yearning for reinstatement. Novelty arises from the fact that no two events are precisely the same, and reinstatement can only be partial. There is a quest to preserve primordial attachments within given rhythmic configurations. Detachment preserves the self-identity and the power of a given event. Re-attachment establishes a kind of security within the entire matrix that nourishes anew that given event. Hence, it provides the conditions for solidifying self-identity, amplifying, intensifying and broadening its powers. A dialectic holds between the two tendencies, a dialectic which sweeps over the entire body and provides for its dynamic equilibrium, a scheme of balances and imbalances.

The overall balance thereby produced is associated with an entirely new set of rhythms that express the interplay of the several factors constituting the body. In addition, these rhythms express the *togetherness* of those factors, and in that sense shape a body-Imago which mirrors to itself the total configuration of the body. The resonance of this body-Imago *is* the Unconscious – i.e., on its lowermost stratum. When bodily activity becomes disrupted, and accordingly disharmonious, the associated Imago registers these imbalances. Hence, the Unconscious is the sphere of registration of body's balances and imbalances [2]. The interplay of both pathologic factors and healthy factors is thus registered as well as the underlying scheme which unites the two sets of factors in a larger configuration. Body's actual condition together with body's strivings toward health are both represented, *re*-presented as an unconscious body-Imago. In particular, the conflict between the two, the actual and the potential, is likewise registered.

Mental Integration: A Dialectic of Regions

With respect to the sphere of mind, analogous considerations hold. Again to cite Peirce:

> Not only do all ideas tend to gravitate toward oblivion, but . . . ideas react upon one another by selective attractions . . . the associations between ideas . . . tend to agglomerate them into single ideas. [They] are attracted to one another by associational habits and dispositions – the former in association by contiguity, and the latter in association by resemblance ([7], 7.533f).

Percepts flow in, interacting with ideas already present on all the different levels of consciousness. Thus consciousness is kaleidoscopic, floating, evanescent. Tensions between its various regions develop. A particularly strong region crystallizes as the self in relation to other weaker regions, turning toward the latter as it inspects, probes, scrutinizes, and controls them. Ideas tend to combine in much the way that events combine. Associated with characteristic resonances, they recognize one another, reject one another, lock in, select, and discriminate idea-trajectories within the field of consciousness. In fact, the more that ideas are associated with earlier ideas of a given kind, the stronger the subsequent bond to like ideas will be. And ideas are mental events, quite analogous to physical events in their capacity for migrations, attractions and repulsions. Within this flowing matrix of ideas, certain invariant themes tend to develop, about which cluster and crystallize certain idea sequences. Those ideas which cannot combine, because the combinations would tend to suck ideas into relatively autonomous but nonetheless powerful configurations within the manifold of consciousness, split that manifold asunder, as it were, into relatively independent regions or foci of activity. The extreme instance of this phenomenon is the radically split consciousness of certain kinds of schizophrenia. A conflict emerges between such pathologic clusterings and the remainder of consciousness, a conflict that hampers the development of the self-center, constricting and deforming that center. Such ideas tend to be relegated to lower regions of the mind, indeed to the oblivion of the Unconscious.

The Pathological Unconscious

In consequence, the pathologic Unconscious consists of diffuse, non-interwoven foci of mental-physical activity. By this activity, the unity of body and mind are diminished; and the Unconscious registers an Imago of this relative disunity. However, in addition, it registers an Imago of the essential unity of body and mind. For the positive potentialities inhering in the relatedness of unselfconsciousness are, once actualized in those spheres, transferred through each stratum of mind, affecting that stratum, to the Unconscious itself. There, both mental unity and the disruption of mental unity blend with analogously imparted unities and disruptions deriving from the body and its embeddedness. In short, the condition of body and mind, in their unity and in the shattering of their unity and in both higher and lower modes of relatedness, all intermix within the Unconscious. The body factors rise to the rhythms of unconsciousness whereas the mental factors fall to those rhythms. In healthy persons, the rhythms are appropriately equilibrated; in unhealthy persons, dis-equilibration prevails.

III. THE UNCONSCIOUS IN COMPORTMENT

According to my argument, every region of human being is suffused with the rhythms of the Unconscious. In turn, these rhythms pervade both the sphere of body and the sphere of mind. In their public aspects, these spheres of activity constitute schemes of spatial compresence – the interwovenness of my body with other bodies and of my mind with other minds – the contemporaneous spread of interwovenness. However, as one passes from either sphere toward the processes which link both spheres, hence mediate their component rhythms – namely the Unconscious – this spatial character diminishes. It gives way to a distinctively temporal character. In effect, the person *'unconscious-es' forth* his mediating rhythms, transmitting their unique and private imprints to body and to mind. Indeed, it as though time were the stuff of which the Unconscious is woven; and it is as though time veritably metamorphoses itself into space. For, in effect, a smooth flowing temporality, in a coherent enmeshing of the movements of memory, is laden with the spatial depositions of contemporary corporeal or mental imprints. By this analogy, the Unconscious is spatialized with pathologic clusterings.

Alienation of Body and Mind

Moreover, only in my Unconscious do I stand alone. In its higher functions, my body is approximately alone; in its lower functions, my mind is approximately alone. For as one approaches the Unconscious from either pole, that of the publicness of unselfconsciousness or that of the publicness of bodies enmeshing with other bodies, aloneness is attained in different gradations. Pervaded by (hence made luminous with) my privacy, both the symbols of my mind and the symbols of my body, my thoughts and my physiognomy, portend the inwardness of my Unconscious, the seat of my creative syntheses. Associated with every set of rhythms on every stratum of human being, whether it be dominantly mental or dominantly physical, are echoing resonances, the reverberations of those rhythms as they sweep over the person. For, there is an element of the physical *and* the mental on *every* level of human being. The lower the stratum, as one descends towards body, the coarser and more discernible are the initiating resonances, and the more evanescent, faint and vanishing are the echoing resonances. On the levels of mind, however, both sets of resonance tend to be subtle, refined and physically quite undetectible – though subjectively quite pronounced. Thus the disparity between the two tends to become greater in body and less in mind. For, in the latter case, as one ascends to the higher strata of mind, they become more and more assimilated to one another.

On the other hand, as one approaches the Unconscious from either side, a frank inversion occurs. In this sense, as one passes toward body, there is a drifting away from reflection proper, and as one passes toward mind, there is a drifting toward reflection proper. In effect, reflection haunts all matter: the more subtle the aura, the more intricately composed the matter. The more intricately composed the matter in certain highly organized configurations, the more mind itself arises. And, always, wherever mind is thus associated with matter, a region exists wherein one may find those peculiar reversals that constitute the processes of unconscious-ing.

The World and Its Influences

According to my view, 'the ultimate seat of every self is the *entire* world' [3, 4]. By this, I mean: the body may be construed as a mosaic of trajectories which conjugate at certain centers, the focal points of reflective activity; and these trajectories extend themselves into the milieu within which the body is embedded. In effect, the cartography of this terrain is mapped onto the body. By a kind of contagion, impulses are transmitted through these trajectories which register themselves 'within' the body as representations of the body's variegated relationships, its schemes of balances and imbalances within its world. By this representation, the reflective body orients itself within that world. Constituted by diverse sensitivities, the person comports himself toward his world in ways determined by the impact of different regions of that world upon those sensitivities. Whether impinging at the mental pole or at the physical pole, such impact reverberates throughout the person – in effect, either ascending to the latter or descending to the former.

Regions in the Self

Moreover, the person may be construed as a composite of different regions of influence, each deriving from a specific mode of locating itself within its local milieu. Composed of many living parts, the human person accordingly represents to himself, however buried their representation be, the specific modes of dwelling in their respective locales of every region of his being. In effect, the representation he frames is constituted by Imagos derived from the mirrorings to itself of the variegated schemes of location, orientation and migration associated with these parts. Hence, in a spatial sense, his being actually spreads itself out over a terrain of considerably greater extent and variety than that of which he is explicitly aware. He frames for himself a cartography of multiple locales, all overlapping yet each distinctive. Moreover, from a temporal point of view, the person inherits the cumulative evolving perspectives contributed by the diverse phases of his own development, a heritage transmitted through the successive metamorphoses he undergoes as his personhood germinates and, in stages, is brought to fruition. In the integrated person, the several contributions made by temporal antecedents and spatially compresent parts all smoothly and, as it were, luminously flow into the coherent manifold of his being.

CONCLUSION

To conclude, if the aura of mental resonances may be regarded as hovering about the body, inscribed as it were upon its constituent rhythms, then every bodily act will manifest, as a symbol portending these resonances, some region of the mental. Every element of comportment constitutes, in effect, an expression of that region; beyond that, the configuration of such elements in their dynamic interplay not merely expresses but actually quite directly *presents* the mental. In this context, 'unconscious' identifies the scheme of rhythms woven at one and the same time of both (physical) symbols and (mental) symbolized; the former in its ineffable components and the latter in its latent content. For, as I have claimed, the Unconscious mediates body and mind; accordingly, it partakes of the character of each. With respect to textural detail, the physical expression, hence presentation *par excellence,* of the mental is of course language. Here fineness of articulation and subtlety of nuance achieve their consummate shape. Yet, in this active sense as utterance, language is but the focal presentation of the mental within the more inclusive context of comportment in general. And the Unconscious informs comportment with rhythms that constantly intertranslate, each into the other, the two sets of resonances of which comportment is woven: the physical proper and the mental proper. When the Unconscious works in unimpeded fashion, comportment becomes, in sum, integrated, smooth, flowing, coherent and luminous.

Fordham University,
Bronx, New York

NOTES

[1] Upon the request of the editors, this paper is truncated from its original. Partly, this was done by myself; partly, it was done by others. Hence, the article is but a vague hint of a proposed new way of looking at mind and body. The original version was considerably longer, and conceptually much more articulated. I consented to present this brief essay as a reminder, a memento as it were, to those who responded enthusiastically to much of the original and as of possible interest to those who enjoy seeking to glean an insight from the bud of an idea.

[2] See my discussion in [1].

[3] See my discussion of mind and body in [2] and [5].

[4] See discussion of archetypes in C. G. Jung [6].

[5] Throughout my account, I lean heavily on Alfred North Whitehead's writings on mind and body. See especially [8].

BIBLIOGRAPHY

1. Feldstein, L. C.: 1975, 'Bifurcated Psyche and Social Self: Implications of Freud's Theory of the Unconscious', in R. J. Roth (ed.), *Person and Community,* Fordham, New York, pp. 43–62.

2. Feldstein, L. C.: 1976, 'The Human Body as Rhythm and Symbol: A Study in Practical Hermeneutics', *Journal of Medicine and Philosophy* **1**, 136–161.
3. Feldstein, L. C.: 1976, 'Personal Freedom: The Dialectics of Self-Possesion', in R. Johann (ed.), *Freedom and Value*, Fordham, New York.
4. Feldstein, L. C.: 1969, 'Reflections on the Ontology of the Person', *International Philosophical Quarterly* **9**, 313–341.
5. Feldstein, L. C.: 1961, 'Towards a Concept of Integrity', *Annals of Psychotherapy*, Monography Nos. 3 and 4.
6. Jung, C. G.: 1959, *Archetypes and the Collective Unconscious*, Pantheon Books, New York.
7. Peirce, C. S.: 1958, *Collected Papers*, Harvard, Cambridge, Mass.
8. Whitehead, A. N.: 1929, *Process and Reality*, Macmillan, New York.

CORINNA DELKESKAMP

BODY, MIND, AND CONDITIONS OF NOVELTY: SOME REMARKS ON LEONARD C. FELDSTEIN'S LUMINOSITY

Philosophy, as the love of wisdom, may also be described as the art of failing. One may fail in this endeavor by loving either too little, or too much. From the former spring neatly reasoned papers, at best: carefully defended and prudently restricted to manageable insights, yet not overly inspiring. The latter gives rise to monstrous designs of a daring imagination, ideally conceived, yet somewhat cloudy and hard to make out. It is customary to excuse the first as publishable samples of scholarly craftmanship and to deride the second as nonsense. But one may wonder whether wisdom, should it ever yield to a wooer, would not prefer the ardent – even though imprudent – lover to the sober-hearted candidate.

Few will pause in assigning Leonard C. Feldstein's essay to the difficult category. He proposes to 'illuminate the role of the Unconscious in the actions of integrated persons' ([4], p. 177). Yet it is hard to decide what he means when he characterizes this role as 'glowing' through actions, qualifying them as 'luminous.' Such luminosity is said to result from a mediation between bodily and mental activity. But whether this mediation results from the simple (spatial) interposition of the unconscious as a medium (like a filter through which the respective activities may pass), or from a 'transformation' in the structuralist sense of a 'rule,' understood as a theoretical construct, remains open. Neither possibility seems compatible with the rhythmic account ([4], pp. 178–179) of the differences between body, unconsciousness and consciousness – whether these differences are meant to arise from a decreasing coarseness on the way 'up' to the 'refined' vibrations constituting mentation ([4], p. 178), or from an increase in orchestration and cohesion ([4], p. 181) as these vibrations ascend to the mind. If all three levels are distinguished by some mode of the basic rhythm (and leaving aside the question of who does the 'rhythming,' since nothing substantial or corpuscular remains), then the unconscious can no longer 'mediate' independent processes ([4], p. 179). It must, rather, itself consist of the middle frequency (or middle-cohesion) of such a range of processes. The unconscious has, so to speak, been transformed from the medium into the message. Yet Professor Feldstein's account of disease ([4], p. 182) again implies that the unconscious is a medium, for it must have the capability of either blocking or

H. T. Engelhardt, Jr. and S. F. Spicker (eds.), Mental Health: Philosophical Perspectives, 191–198.

freely transmitting processes from the outer strata (i.e., the body and the conscious).

The rhythmic model also makes it impossible to account for the meaning given to body. Within the body-level, one stratum is determined as 'function' ([4], p. 179). Yet the concept of a function is of a different logical order than that of a (patterned or structured) vibration. While the latter is descriptive of a thing apart from purposes, the former envisages something beyond that thing in its present state, towards which it is supposed to be striving – or beyond that thing itself, in the service of which it acts. This would be similar to encountering a stuttering rhythm, and then characterizing it as one 'aspiring' to be a Vienna waltz. And even if prolonged observation would confirm that the rhythm approximates the waltz and finally settles down into one, this would – on purely rhythmic grounds – not warrant the conclusion that being a waltz had been its 'function'. Again, the 'liquefying' ([4], p. 180) of structures into 'interwoven paths' leaves us only with structures, since these paths must in turn be conceived as 'structured' vibrations. Moreover, how functions should be 'inscribed' upon body ([4], p. 178), when they had just contributed a stratum of that very body itself, and how they in addition should count as 'feelings,' or something belonging to consciousness, rather than first connecting with the adjacent unconscious, is difficult to apprehend.

The unconscious is said to function differently in health and disease, both kinds of functions (a) *revealing themselves* in a person's behavior. Yet at the same time they (b) *condition* 'man's characteristic searchings.' These searchings, however, would seem to form part of a person's behavior as well. Hence, such behavior must at the same time (a) *signify* as a symptom the functioning of an otherwise inaccessible unconscious and (b) *be explainable* on the basis of such functioning. Hence, if that functioning is revealed only in behavior, behavior must explain itself.

Of course, one may conclude in any event that the functioning of the unconscious (supposing it to be understood as a medium for transmitting processes) is the decisive determining factor for health or disease. But even then it is not at all clear how this would account for diseases apparently originating in the body. In sum, then, the role of the unconscious cannot be illuminated as long as the concept itself remains obscure.

But it seems that Professor Feldstein's essay can be better understood if an underlying motivation is spelled out. An account of man as a conceptual amphibian, consisting of matter and spirit, is certainly not satisfactory in view of the unity some humans believe themselves to possess. Yet the terminology of this dualism pervades the literature of psychiatry, psychology, philosophy, and of some of the medical disciplines. Though the inadequacy of this model is generally acknowlecged, few have made fruitful efforts to overcome it. The author, then, engaged in establishing a perspective with reference to which the unity of man can be exhibited. His attention to health and disease can be interpreted in the light of an endeavor to grasp the manner in which bodily diseases can be said to affect consciousness, and vice versa. His investigation of the unconscious is not meant as an analysis of facts, but rather as a presentation of a conceptual framework within which to reconcile seemingly incompatible substances. His emphasis on consciousness *as* body ([4], p. 177), or grounded *by* body ([4], p. 178), as well as his insistence on body as 'enveloped' by consciousness, signal the challenge to think the two into one. His attempt to spell out this unity by means of the available terminology of depth psychology then presents a fascinating conceptual experiment.

Such redesigning of the interior of a conceptual scheme (i.e., the categorial system of body-mind relations), requires, however, that some basic and well-established framework (i.e., the terminology of depth psychology) is left intact. While it would indeed be unfair to refuse agreement after merely surveying the decorative items (or to judge 'luminosity' and 'rhythms' in isolation and before they have been assigned that proper place at which they can contribute to the coherence of the whole) – one may justly object if Feldstein tries to nail his new pictures to a wall he himself tore down, or attach the real attributes (of body-mind interaction) to a 'universal "wallhood",' (or to a concept of the unconscious which could be particularly defined only if those attributes could be made to stick where that lofty abstraction provides no support).

Let me paraphrase this image. Feldstein has set for himself the important task of transforming what is given as isolated data in the medical, psychiatric, and psychological sciences, and in the philosophy of mind, into one homogeneous whole. This can be compared to Ovid's attempt to transform the several isolated stories of Greek mythology into the unity of a Latin poem, or to Berkeley's task in his *Siris* to transform the spatial manifold of isolated world-objects into ideas that are unified by the eternal Now of God's mind, or to Hegel's endeavor to transform the multiplicity of concepts furnished by the tradition of philosophy into one comprehensive system. In each of these cases the transformation to be effected is also thematized: in Ovid by the metamorphoses of humans, spirits, and gods into stones, rivers, plants, and animals; in Berkeley by the hierarchical order of levels of being, with each more essential nature causing what appears as effects in the adjacent grosser medium; for Hegel by the concepts of being-in-itself, being-for-itself, and being-in-and-for-itself. Analogously Professor Feldstein thematizes the desired transformation as 'luminosity' in the sense of a transparence of one level of human being to the other, or of the capacity of each level to 'shine through' into the next. A similar effect is intended with his 'doctrine of vibrations,' depicting a constant exchange of momentum between the coarser rhythms of the body and the finer ones of the mind.

Yet in the first three examples, the transformation, which had been both intended and thematized, is in addition realized, or at least exemplified, through the thematizing discourse itself. Thus, in Ovid's poem, the 'metamorphoses' are not only signified or reported, but 'enacted' through the dramatic tension of his verse. The mythological figures, while being described as changing into natural phenomena, are molded into the artistry of words and into the metaphoric network of a poetic imagination. Similarly, in Berkeley, the chain of reflections along the search for causes and principles structuring his presentation reflects the ontological chain of beings represented. When finally those orders of existent things have been converted into orders of ideas in the mind of God, the reader, who began by representing these various beings in his own ideas, is in turn impelled to reconsidering the chain of these ideas as, already, constituting those beings themselves.[2] And, in like manner, Hegel's dialectical arrangement of subject matter exemplifies on the methodical level what his subject matter is about. The proposed unification of comprehension and of what is comprehended is realized as the discourse proceeds to constitute that very subject matter which it speaks about, and to elicit in the reader an understanding of that subject, as it arises from his own progress in understanding. Feldstein's presentation, in contrast, is not restrained by having to exemplify its own assertions. As the order of his subjects lacks necessity, and as the sequence of his statements can often be reversed, these statements fail to form links in a chain of reasoning each of which is indispensible for establishing the conclusion. By not rigging the display of his arguments in such a way that the act of 'having followed them' already constitutes a tacit commitment to their proposed intelligibility, he does not compel the reader's assent.

To be sure, the presentational realization of representational content is of value only if it serves not merely as a presentational device, but, in addition, as a hermeneutic tool. Only then is the inner methodical coherence of a certain order of proceeding not just imposed, but, rather, discovered in the given material. Thus, that material appears, after the fact, to have on its own 'lent itself' to just such an account. A method must have 'been called for' in order to do justice to the subjects at hand, if the intelligibility granted to a novel approach is to appear *relevant*. We are not convinced of the value of Hegel's account on the basis of his chapters on the Middle Ages in his *History of Philosophy*, where he forces a superficially assessed tradition into his preconceived framework. But the discussions in the *Phenomenology* account for the reflexive relation between – say – master and slave [5] in what suddenly appears to be the only adequate manner for doing so. Furthermore, this account by Hegel, which at first seemed only an explication of previously neglected implications, ends by making us believe that even the phenomena of self-consciousness had always somehow belonged to that subject matter. Hence, what we would have never suspected, comes to compel us to understand anew. Novelty is acceptable only when it accounts for accepted knowledge, in which traditional concepts have acquired a specificity that can impose restrictions on their all too arbitrary employment in novel formulations.

Professor Feldstein's interpretation of the unconscious, of mentation, of body and association does not account for the systematic complexity and differentiation of meaning which these concepts have accumulated through their function in various explanatory models. Thus, Feldstein does not appeal to intersubjectively acknowledged standards of reality which could either resist or favor his claims. His concepts are so easily available to him that they do not force him to explain in detail how their meaning differs from the established ones. As a consequence, once the interpretation of man's unity has been presented, the constituent parts, as they have not retained a conceptual life of their own, become obsolete. The *explanans* no longer requires the *explanandum* and is thus in turn rendered vacuous. If Feldstein's levels or strata of the being of humans (i.e., the rhythms of consciousness, of the unconscious, and of the body) are 'really' distinguished only by degrees of rhythms, then how could we ever have been so deceived as to call them by separate names, understand them by disparate definitions, and assess them in radically different manners?

In the process of overcoming a state of separation, or of alienation, that state should not be simply annihilated, lest one wonder whether anything has been overcome after all. There must still be a place for the problem in the solution. Though Ovid wanted to overcome the separation between poetry as a means of presentation and mythology as a subject matter to be represented, he did not simply stipulate a newly invented 'poetical mythology' as the desired unification. Instead, he took up the available tradition of mythological stories by exhibiting their transformation into his poetic vision of an animated nature. In this process, the original cause of 'alienation' was overcome, but only by being re-enacted: Just as the translation of Greek sources into the Latin language had occasioned the discrepancy, so yet another translation of stories into poetic images is shown to re-establish the unity between content and form. Similarly, Berkeley, in his endeavor to overcome the alienation of man from God, does not just confront us with a mystical wedding, but begins with the everyday concerns of health and disease, and with the explanatory models of the science of his time. It is by penetrating these with an ever-increasing precision of analysis that he uncovers the result (his solution) as having already been present in his point of departure (in the state of affairs constituting his problem). By gradually disassembling the conceptual barriers which the philosophical tradition has accumulated, he recovers the human condition prior to the process of that alienation while at the same time showing the necessity of that process which had begun with the fall of man. Hegel as well, in the *Phenomenology*, starts with the common presumption that the 'here' and 'now' are the bases for all our certainty, and compels us, by a stepwise refutation of all further such prejudices, to accept his results. However, the necessity of having these prejudices so that they can count as points of departure is then reconstructed with negation as the condition for synthesis and truth. In each of these cases, the conditions for the problem are justified as well as overcome in the final unifying vision, thus certifying the *very possibility* of the accomplished task.

These metatheoretical considerations concerning the intelligibility, relevance and possibility of novelty are crucial in the cases described, because (if we disregard the instance of poetry where such reflexivity is a constitutive element) they all concern the thinking of something unthinkable: namely, of the unity of real and intentional existence (or, in an old-fashioned vernacular: the unity of *esse extra* and *esse in intellectu*). This unity, in the case of Berkeley, is unmasked as a divine mind containing (representing) the world 'behind' that divinely fiery spirit which is in turn contained by the (represented) world. For Hegel, this unity is established as the 'transition' from objects as conceived (represented) to objects as concepts (manners of representing). Professor Feldstein, when unifying spirit and matter as components of man, is in particular obliged to unify body and mind, or that which is capable of signifying (or presenting) only itself to others and that which is capable of signifying (or presenting) others to itself. By considering the body not only as present in mental awareness ([4], p. 177) but as 'pervading' it, or by suggesting that unrecallable ideas may 'sink' ([4], p. 178) into body, Feldstein is again committed to reify information, or to intentionalize real objects. Any such endeavor is necessarily incompatible with common sense or with our scientific terminology. Thus we are left with a novel language requiring that rational conditions for novelty be observed. Accordingly, for Berkeley that 'behind', and for Hegel that 'transition', depict a kind of mysterious gap in their reasonings, yet one around which rational arguments have been arranged in such a manner that we are persuaded to tolerate the former for the sake of enlightenment received by the latter. Once mystery is clearly localized and thereby admitted as a condition for the discourse, we are no longer mystified by it.

It is such conceptual confinement of wonder which is missing in Professor Feldstein's account. The undomesticated novelty of his formulations renders his theory something like a 'Yaqui way of psychiatry' [1]. It is not fair to judge it by the criteria of standard reality. But a lot of work is required to make a non-standard reality real. Still, his essay is helpful because it delineates a worthwhile direction into which such work may be channelled. Leonard Feldstein's essay has to be seen in the context of a larger theoretical framework, which he is elaborating into a series of books. We can expect that this work will fulfill the promise given here.

The Pennsylvania State University,
University Park, Pennsylvania

NOTES

[1] I would like to thank Tristram and Susan Engelhardt for linguistic and stylistic improvements, and improvements on those improvements respectively, and Ruth Walker for helpful criticism.

[2] A more extensive discussion of this aspect of Berkeley's *Siris* can be found in Corinna Delkeskamp, 'Temporalité et Aliénation chez George Berkeley' [2].

[3] Such a gap is most clearly exemplified in Plato's dialogue *Timaeus* (35a). There, in a slightly different way, the soul not only *partakes*, as an intermediate *existence*, of both what is ideal and what is sensible (as a power to move the *sensible* according to *ideal* laws), but also, as *mind*, *pronounces* concerning the sensible world, and objectifies it. This double function (of partaking and pronouncing) of the soul is *implicitly* present already in the first step of establishing its intermediate *existence*: The demiurge is not simply mixing two substances. Rather more precisely, he is combining something that relates to things, with something that relates to what things partake in. Loosely speaking, this is like trying to mix real water with the idea of wine – the result is not watery wine but a logical monster. This monstrosity (a gap) is *explicitly* revealed (and thereby acknowledged) by the odd manner in which the final mixture is effected: In the first step both elements (the divisible – or what is akin to the sensible – and the indivisible – or what is ideal) were brought together to form a monstrous intermediate being. But the second step of mixing that intermediate nature again with each of its original elements amounts to combining a term of the logical type of monsters with two terms of the logical type of non-monsters, which is somewhat more unorthodox than trying to combine terms (say, for the enumeration of objects) of different levels. Thus, on the level of the (mixing) procedure, the 'monstrosity' (the gap) which was only implied in the disparate elements of the first mixture is made explicit. While the logical incompatibility of the elements is blurred in the first mixture, the oddity of the second mixture makes sure that that blurring does not itself get blurred.

Strictly speaking, in the second mixture the original elements (the indivisible, unchangeable, and what is generated and divisible in bodies) are viewed no longer with respect to their existence alone but also with respect to how they appear to human comprehension. The former, now, is the same (i.e., that which always remains, and can therefore be known) and the latter is the different (i.e., that which is subject to change and open only to opinion). Mine is, obviously, an attempt to make sense of Taylor's reading of the text, and to avoid Proclus' solution as recommended by Cornford [3]. Proclus gives only an image of a hierarchy of beings; he avoids the logical problems arising from the type-distinctions among those beings. According to the understanding suggested here the second clause (concerning the second step in the mixture, i.e., of the intermediate nature with each of its original components) is no longer 'a pointless repetition of the first' ([3], p. 61). The identification (if we abstract for the moment from the change from ontological to epistemological language) of the *indivisible* and *divisible* with the *same* and the *different* seems, moreover, justifiable when we consider that it is that very created being (i.e., that which is not always the same with itself but appears ever *different*, and is contrasted with its ideal and unchangeable original [*Timaeus*, 27d 4]), which gets *compounded* (Timaeus, 31c ff) out of elements. It is thus objectively (through change) and conceptually *divisible* into those elements, and is distinguished again from the indivisible original. Even Cornford works towards such an understanding. ([3], p. 65.). Neither is Cornford's reference to the *Sophist* helpful, for, as he himself indicates, the manner in which sameness and difference are treated in both dialogues is very different: in the *Sophist* (254e ff) both are applied (in what amounts to a 'grammer of ideas') to ideal kinds (or forms) alone, presenting a sort of meta-kinds. The problem arises because on the one hand all such kinds remain always the same, thus partaking of sameness (in contrast to the sensible things partaking in those kinds, but being themselves always different); on the other hand even among such kinds one can speak of what they are for themselves (namely: same) and what they are with respect to others (namely: different). In the *Timaeus*, however, sameness and difference are not considered as transcendentals at all, but are allotted to the ideal and the sensible realm respectively. They appear in the *Timaeus* only with respect to the first ('on the one hand –') of the two considerations which are juxtaposed in the *Sophist*).

Thus, the soul in Plato's account is *intended* as a unity of being and thought (or of the act of existing and the act of meaning) similar to that which Professor Feldstein has *in mind*. In the former presentation the unity of *reality* and *meaning* which Plato *meant* to *realize*, is *realized* by means of the peculiar arrangement of the elements in his discourse. The mysterious gap between reality and meaning necessitates an account of the mystery (i.e., if I am still following, an *explicit* gap between what he *meant* and what he was able to realize through his discourse). The oddity of his problem, as the condition of a need for novel language, while arising from a logical gap, gives rise to a methodological gap. It is such a discursive realization of Feldstein's meaning which is lacking in his account. The reader, forced to achieve that realization in his own mind, is both stimulated and puzzled.

BIBLIOGRAPHY

1. Castaneda, C.: 1974, *The Teachings of Don Juan: A Yaqui Way of Knowledge*, Pocket Books, New York.
2. Castelli, E. (ed.): 1975, *Temporalité et Aliénation*, Aubier, Paris.
3. Cornford, F. M.: 1959, *Plato's Cosmology, The Timaeus of Plato*, The Bobbs Merrill Company, Inc., Indianapolis, New York.
4. Feldstein, L.: 1977, 'Luminosity: The Unconscious in the Integrated Person', in this volume, pp. 177-189.
5. Hegel, G. W. F.: 1804, *Phänomenologie des Geistes*, in Hermann Glockner (ed.), *Sämtliche Werke*, Friedrich Frommann Verlag, Stuttgart-Bad Canstatt, 1964, Vol. 2, pp. 153–158.

SECTION IV

ACTING FREELY AND ACTING IN GOOD HEALTH

IRVING THALBERG

MOTIVATIONAL DISTURBANCES AND FREE WILL

I plan to struggle with a conceptual knot which should be both intelligible and challenging to mental health theoreticians as well as to philosophers of action. This may help us achieve additional clarity about what it is for a person to act more or less autonomously, to exercise a greater or lesser degree of command over her or his own behavior. Regarding mental health, I shall inquire: Must a person be of relatively 'sound mind' in order to act freely? Do some (not all) forms of so-called mental illness,[1] including alcoholism and drug addiction, decrease his mastery over what he is doing? These broad questions originate from a fairly definite paradox about motivation and freedom – to which I have heard three promising philosophical answers. The replies show notable affinities with, and stark dissimilarities from, a standard psychoanalytical account of how neuroses and psychoses diminish one's authority over his behavior and moods, and of how therapy might restore it.

I. OUR PARADOX

Crudely and schematically, it begins with the fact that only when a person wants to perform an action of kind A, and does so, can we say that he acted freely. We would not characterize him as doing an A-type action unwillingly – for example, under coercion – unless we assumed that he is against doing so. Someone altogether bereft of pro- and con-attitudes, even lacking a distaste for being bullied and threatened, would not be said to act either voluntarily or unfreely. So far no riddles. But shall we say a person acts willingly or unwillingly if he is driven by the obsessions, delusions, and phobias that disturbed people frequently act on – and which impel us all at times? The anomaly here is that those very circumstances which make free action possible, wanting to do something and doing it, now seem to undermine liberty. Should we say that such people do not act freely because the motives which stir them are not really theirs?

Freud saw the neurotic as afflicted with 'ideas and impulses' which 'give the patient himself the impression of being all-powerful guests from an alien world' ([9], 1916–17, XVI, p. 378). Freud tells the sufferer: 'Part of the activity of your own mind has been withdrawn from your knowledge and

H. T. Engelhardt, Jr. and S. F. Spicker (eds.), Mental Health: Philosophical Perspectives, 201–220.

from the command of your will . . . [Y]our . . . instincts have rebelled . . . [and] extorted their rights in a manner that you cannot sanction' ([9], 1917a, XVII, pp. 141f). But Freud's political metaphors do not enable us to separate (1) attitudes which really belong to us, and (2) rival 'ideas and impulses' which supposedly overpower our 'own' attitudes, and usurp control of our behavior. After I examine the elements of our problem more systematically, I shall deny that either psychoanalysis, or the philosophical answers in vogue, will cogently explain the distinction we seek. Nevertheless I agree with one contributor that there must 'be a way of saying that some desires are more intimately related to the self – are more its own' ([12], p. 43); but I doubt that I can map the 'way.'

II. DOING WHAT ONE WANTS

Imagine that I have successfully performed some individual action *d*, which is of type *A*. I have done so because I wanted to execute an action of type *A*. For instance, I jogged over a ridge because I wanted to engage in rapid uphill and downhill ambulation. I harp on 'want' and similar verbs of 'conation' only for convenience. Our puzzle about freedom and motivation crops up just as insistently if we assume that I performed *d* because of my calculations, my beliefs, my mood, my proclivities, my current projects, my ambitions, my pretensions, the promptings of my conscience, my sense of honor, decorum or personal obligation – if these fit the context. What matters is that the chief impetus to action was my current attitude. As opposed to what? The fundamental contrasts are with irresistible natural forces; laws backed by probable sanctions; duress and threats. This basic but surprisingly problematical comparison will occupy the next two sections.

Before that, a word on 'because.' Must this conjunction mark a cause-effect relationship between separable happenings, one either 'mental' or neural, the other a sequence of limb movements? We should not rekindle debate over the causal or non-causal status of motivational factors.[2] All we need assume is that my attitude was crucial to my behavior under prevailing conditions. Perhaps 'crucial' in the sense that, if I had *not* wanted to do something of kind *A*, then I would not have carried out *d*, or any other *A*-ish deed, as conditions stood. For instance, no as yet unseen brush fire, or trigger-happy owner of the hill, was lurking backstage, ready to compel me to perform another, slightly different *A*-type action, *d′*, in case I had omitted *d*. When such 'potential constraint' is present, though unsuspected by me, the chief impetus for *d* is still my wanting to perform an *A*-type act. But it would be an overstatement to insist that this desire is the main factor which accounts for my carrying out an *A*-type action. At least we should acknowledge that it was 'overdetermined' that I would do something *A*-ish. Unfortunately we cannot digress to settle the raging controversy whether I perform *d* freely, if I act in ignorance of such potential compulsion.

III. OUTRIGHT CONSTRAINT

It is difficult enough to analyze overt, 'physical' compulsion, which presumably contrasts with acting because one wants to. A renowned action theorist offers this illustration: 'If a hurricane wind blows you twenty yards across a street you cannot be said to have crossed the street voluntarily, since you were compelled to do it and given no choice . . . ' ([4], pp. 274f). Does he mean that you were lifted off your feet and deposited twenty yards away? Then your change of location was not your doing, and therefore not anything 'you were compelled to do.' On the contrary. Something happened to you: you were carried by the gale. I would propose a similar analysis if burly gangsters raised you into the air and transported you twenty yards.

To get *bona fide* instances of being constrained to act, we must alter our twin stories, as follows: (i) You realize that the wind is so violent that if you do not walk with it across the street, it will shove you there anyway; so you reluctantly stagger in the direction it is buffeting you. (ii) The abductors rough you up, and order you to accompany them; convinced that if you refuse, they will simply drag you, you resist a bit, and being shoved occasionally for good measure, you comply.

IV. DOES ONE WANT TO PERFORM THE ACTION HE IS COMPELLED TO PERFORM?

Evidently there are grim differences between these examples of constraint and our previous model of jogging over a hill because one wants to. Yet we see a deeply baffling similarity: not only do you act, instead of having something happen to you, when you are compelled; but also in some sense you act because you want to – while nevertheless somehow wanting *not* to act as you must. If we can understand this 'wanting to and wanting not to,' it may help us solve our major puzzle about how disturbed people act against their will.

As a start, what separates my hill-climbing because I wanted to, and your crossing the street because you wanted to? One disanalogy is that if I had not wanted to climb, I would have remained below. But you would have left this side of the street even if you had wanted to stay put. You would not have performed the act of ambulation which we imagined, namely, *e*. However, the gunmen or the wind would have forced you sooner or later to perform a slightly dissimilar act, e'; or else they would have propelled you. Yet apparently you wanted – really wanted, wanted more strongly – to remain where you were. How shall we explicate these contrary wants?

V. A HIERARCHICAL ANALYSIS: THE COERCED INDIVIDUAL DESIRES NOT TO DESIRE TO SUBMIT

One currently favored account of coercion takes what I call 'hierarchical' form. Professor Gerald Dworkin, for instance, considers the victim of a highwayman, who unwillingly surrenders his wallet. 'What he doesn't want to do,' says Dworkin, 'is to hand money over in these circumstances, for these reasons' ([2], p. 372). Dworkin thinks our aversion is directed toward our motives:

> Men resent acting for certain reasons; they would not choose to be motivated in certain ways. They mind acting simply in order . . . to avoid unpleasant consequences . . . . Part of the human personality . . . takes up an 'attitude' toward the reasons, desires and motives which determine . . . conduct . . . . We consider ourselves compelled because we find it painful to act for these reasons ([2], pp. 377ff).

The upshot is that Dworkin boldly defines 'acting freely' as acting 'for reasons [one] doesn't mind acting from' ([2], p. 381).

Professor Harry Frankfurt elaborates this split-level strategy. First he announces that 'one essential difference between persons and other creatures' is that persons 'are able to form . . . "second-order desires"' ([6], p. 6). Next Frankfurt coins the term 'second-order volition' to label one's higher hankering that a particular ground-floor inclination 'be the desire that moves him effectively to act' ([6], p. 10). Frankfurt then tightens his requirement for personhood: 'It is having second-order volitions, and not having second-order desires generally, that I regard as essential . . . .' In his lexicon, 'wantons' fall short. These

> agents . . . have first-order desires but . . . are not persons because, whether or not they have desires of the second order, they have no second-order volitions . . . . A wanton . . . does not care about his will . . . . He does not care which of his inclinations is the strongest ([6], p. 11)

In later essays, Frankfurt applies this schema to coercion. He thinks

> an offer is coercive . . . when the person is moved into compliance by a desire . . . which he would overcome if he could . . . a desire by which he does not want to be driven . . . .

A person's autonomy may be violated by a threat in the same way .... In submitting to a threat, a person invariably does something which he does not really want to do ([7], pp. 80f).

Most recently, Frankfurt has demarcated two cases of doing what one 'does not really want to' under compulsion: (a) The unappealing 'alternatives' which a person confronted were 'a set from which he did not want to have to choose;' nevertheless he 'preferred without reservation the one he pursued' ([8], p. 114). (b) The agent's 'inclination to avoid the undesirable consequences he faces is irresistible' ([8], p. 116). Frankfurt's hierarchical interpretation is that in case (a) the agent's 'action is in accordance with a second-order volition,' while in (b) he 'is moved to act without the concurrence of a second-order volition' ([8], p. 116). It is unclear whether Frankfurt thinks the person in (a) acts freely or unfreely; but Frankfurt judges him 'morally responsible' ([8], p. 117). So I shall dwell on (b), and Frankfurt's remarks about the coerced individual who would like to 'overcome' his first-order desire, because it 'is one by which he would prefer not to be moved' ([7], p. 82).

My first objection against the hierarchical accounts of Frankfurt and Dworkin is that the ascent looks unnecessary. What is the holdup victim more opposed to: being deprived of his money, or abandoning it 'for these reasons'? Surely the former! Again, is it not less 'painful to act for these reasons' than to lose one's billfold? Don't we 'mind acting from' the desire 'to avoid unpleasant consequences' much less than we mind the consequences themselves? Moreover, why should Frankfurt's recipient of coercive threats want to 'overcome' his effective desire? Why should he, in case (b), withhold second-order approval of it? As far as I can tell, only because the actions to which it drives him are somehow intrinsically bad, or because they produce unwelcome results. Thus it seems gratuitous to suppose a constrained person cares mainly about his will.

My grumble is only that in Dworkin's and Frankfurt's hand-picked examples of coercion, it does not seem to be true that the victim's dislike is principally aimed at the reasons or motives on which he acts. I need not maintain that in every imaginable case a person cares exclusively, or even primarily, about the action he is compelled to perform. But if you look without preconceptions at alleged instances of second-level desire and volition, my hunch is that you will usually reinterpret them as directed toward acts or consequences of action. Take the examples Dworkin wishes to contrast with the highwayman's victim, who 'doesn't want ... to hand money over ... for these reasons.' Dworkin imagines someone who instead 'might want to hand over some money . . . because he is asked by a relative, or because he is feeling charitable, or because he desires to rid himself of wordly things' ([2], p. 371). Are these really cases of second-order wanting to be moved by the impulse to help one's kin, the urge to help all people, and the desire to be unburdened of wealth and possessions? Doesn't the person mainly approve of the corresponding actions, and isn't that why he wants to be thus motivated?

I have the same suspicion about a parallel case invented by Frankfurt:

> A man . . . decides to . . . give the money in his pocket to the first person he meets . . . The first person he meets points a pistol at his head and threatens to kill him unless he hands over his money. The man is terrified, . . . and . . . hands over his money in order to escape death . . . . [If the man had] handed over his money with his original benevolent intention . . . [then] he would not have been coerced in doing so. His motive in acting would have been just the motive from which he wanted to act ([7], p. 82).

Of course nobody likes to experience terror; but that is beside the point. Isn't it more plausible to hold that Frankfurt's philanthropist would rather not engage in the action of giving-under-threats, and would rather engage in some giving-without-threats? The latter kind of giving might appeal to him because it is safer and pleasanter.

Besides objecting, as I have, that the double-decker analysis of being forced to act is unnecessary, I have a more serious complaint: it fails to make sense of our assumption that compelled people 'really,' more strongly, or most want to act differently from the way they are constrained to act. Such comparative phrases imply that a compelled individual's second-story aversion is more potent or intense than his ground-floor desire to submit. But then shouldn't we expect it to prevail over its lowly cousin? Hierarchical views therefore seem at odds with the very fact which needed explaining: the fact that one's desire to comply 'wins out.' How is this possible if one's upper-floor disapproval is more genuine, powerful, or authoritative? So a hierarchical theory does not seem to illuminate compulsion.

VI. A SIMPLER ANALYSIS OF CONSTRAINT

We will give hierarchical views another run when we get to wayward motives and 'inner compulsion.' We still have to demystify the more fundamental contrast between just doing what one wants, and the peculiar kind of wanting that is involved when a threatened person chooses the least dangerous or baneful course open to him. I think we can get by without the apparatus of second-order desires and volitions. To prepare the ground, however, I must digress briefly on how people use the key verb 'want' and its equivalents.

When we know, or anyway justifiably believe, that in our circumstances it is going to be impossible – empirically, legally, morally, even logically – for us to do something of type A, we almost never say that we want to. Usually we declare that we wish we could, that we long to, or that we would give anything to be able to perform an A-type action. Normally we reserve 'want' for projects that we have no reason to think ourselves unable to execute. There are exceptions to this rule of usage: principally extreme settings, where the agent has nothing to lose, whatever happens, and also tries to do something A-ish. For example, 'lifers' at the Alcatraz Island penitentiary occasionally attempted to scale its walls, and to swim across San Francisco Bay until they reached the mainland. Apparently none succeeded. It would sound grotesque to deny that they wanted to break out – and pedantic to insist that they only wished or hoped to. But in this special context the person is acting on his want, taking some measures that generally have some chance of leading to the kind of result which presently looks unattainable. This is a far cry from our earlier examples of you being forced by hurricane winds or by gangsters to cross the street. Perhaps you were so frightened, believing that you had a lot to lose, that you made no full-fledged attempt to remain where you were. In that case we have no analogue to the prisoner's abortive escape, by reference to which we can say that you wanted – 'really' or more strongly – not to cross the street.

How then shall we convey our stubborn hunch that this was your attitude? Our brief analysis of how people deploy the verb 'want' may resolve this impasse over your attitude against doing what you are compelled to do. We admit that you wanted to do the least of evils, and withdraw the commonsense claim that you really desired not to. Then we add that your top preference would be for something quite different. Had you known of it beforehand, you would have stayed clear of the situation. And now that you have blundered into it, if you were asked to rank resistance, compliance and escaping to the *status quo ante,* or simply escaping, you would choose some form of evasion ([13], pp. 461f). You might be able to specify in positive terms what circumstances you would prefer. Perhaps you could even delineate the sorts of actions you would prefer to be doing, if you were miraculously delivered from the constraining situation. But I emphasize the negative point, that above all you dislike being in these restrictive circumstances. Now if I am jogging over a hillside because I want to, of course I can always dream up more ideal conditions. But usually I would not choose to return to the pre-jogging scene. Certainly I would not rank simply being out of the jogging situation as preferable to my current activity.

This is not an analysis of coercion. It does not set forth either logically sufficient or logically necessary conditions for saying someone is forced to perform an action of type *A*. Rather, I have characterized the pro- and con-attitudes of the constrained individual, which were sources of our puzzlement. Evidently it is a simpler characterization than the hierarchical account. It also yields a dividend: it does justice to the suggestions of Dworkin and Frankfurt that their protagonists did not 'want . . . to hand money over in these circumstances.' and 'did not want to have to choose' among disagreeable alternatives. It also articulates our belief that coerced people have both a pro- and a con-attitude. Most notably, it re-establishes the contrast between being compelled and doing what one wants. Only the constrained individual always prefers returning to the previous situation, or simply getting out of his present circumstances altogether. Having achieved this much insight, we can investigate analogies and dissimilarities between constraint and motivational disturbances.

VII. DESIRES WHICH RESULT FROM HYPNOSIS AND SECRET MEDDLING

I'll approach our main comparison by way of two bridging examples. Opponents of soft determinism regularly parade these out as counter-instances to the soft determinist principle that whenever you do what you want, or what you have chosen, you act freely. Among others, Professor Flew seemed to endorse that principle, citing as a paradigm of free action the marriage of two yound people, Murdo [*sic*] and Mairi [*sic*]. They deliberate carefully, finally selecting each other. They would not have married if they had chosen not to. Finally, they they are not under any pressure – from their families or whatever ([5], pp. 149–53). Attacking Flew, Professor MacIntyre embellishes his narrative, in order to refute the soft determinist assumption that choice brings liberty. MacIntyre's supplementation reads:

> [Murdo] has visited a hypnotist who has [against Murdo's will] successfully made three suggestions to him . . . : that he shall consider the merits of his female acquaintances, that after due reflection he shall choose [Mairi] . . . , and that he shall forget his visit to the hypnotist . . . . [Murdo's] decision is . . . a free act on all Flew's criteria. He considered the possible alternatives . . . ; and if he had chosen otherwise, he would have been able, in Flew's sense, to implement his decision ([11], p. 244).

Has MacIntyre demolished Flew's criteria – not to mention soft determinism generally? Our account of compulsion equips to parry this counter-example. The mesmerist's interference provoked Murdo's decision just as inexorably as overt bullying and observable natural forces engendered our street-crosser's desire. Murdo has even less control of his behavior, since he is unaware of the principal influence upon his motivation, and unable to combat it. Furthermore, his sudden matrimonial bent, particularly toward Mairi, does not spring from his evolving system of beliefs, goals, moods – not even from any neurotic impulses which afflict him. The hypnotist imposed it; and it may well clash with his dominant interests. It is virtually the hypnotist's, not Murdo's, inclination. So while this is a case of unfree behavior, it is not an instance of the agent doing what he wants.

I would deal similarly with the hallowed fantasy of the 'Devil/neurologist,' which figures in a symposium between Professors Don Locke and Harry Frankfurt ([8], pp. 104ff, 120ff). This has several variants, not all recorded by Locke and Frankfurt: (1) A Devil, or a diabolical neurologist ('D/n' for short) directly and covertly manipulates all of his victim's (V_1's) attitudes, perhaps by constant tinkering with V_1's brain. (2) D/n only occasionally, but quite unpredictably, does this to V_2. (3) D/n creates V_3 from scratch, inalterably programmed to display just one set of cognitive, affective, and conative responses forever, regardless of what 'reinforcing' or 'aversive' experiences he meets with. No subsequent manipulations by D/n are necessary. (4) D/n wires V_4 at 'birth' with some fixed basic drives, emotional dispostions, patterns of reasoning, and Cartesian–Chomskian 'innate ideas.' D/n also endows V_4 with numerous open capacities, to be developed or inhibited by whatever learning experiences he has. Beyond this, D/n leaves V_4 alone. (5) The same setup as in (4), except that D/n makes V_5's 'innate' core of attitudes and ideas flexible – modifiable by learning.

As Frankfurt observes, the puppet we labeled V_1 should not qualify as a genuine person, having thoughts, desires, and so on – however crudely or subtly we define personhood. Consequently we cannot suppose that V_1 is a person who is doing what he wants, but unfreely. V_2, on the other hand, is a person with attitudes of his own – but only while D/n stays away. The attitudes inflicted upon V_2 by D/n are more like D/n's. When V_2 acts because of them, he is not doing what he wants, and so the question whether he acts freely remains tabled. V_2 is in the same boat as MacIntyre's hypnotized bridegroom. As to the totally and rigidly programmed V_3, D/n has visited the equivalent of lifelong hypnosis upon him. We cannot speak of 'what V_3 himself wants to do.' V_4 and V_5, however, strike me as borderline and clear instances, respectively, of someone doing what he pleases. Merely because D/n outfits them with inalterable or with malleable attitudes, we cannot conclude that they lack attitudes of their own, or that they act unfreely

– even when they act on the D/n-given attitudes. Case (4) seems undecidable, until we discover how 'dominant' or 'central' the attitudes are which D/n implanted in V_4, and how these fit in with the outlook which V_4 has managed to develop. But D/n is surely innocent of gross meddling with V_5's motivation. Hence V_5 is capable of doing what he wants – and freely. If V_5 grows up to be an arch-fiend, or a benefactor of mankind, D/n probably deserves some of the blame or credit. However, I am deliberately circumnavigating problems of responsibility.

I do intend to see how these bridging examples link up with our main contrasting pair: opting for the lesser evil, under compulsion; and doing what one wants. The person forced to cross a street resembles our hypnotized bridegroom as well as V_2 and the non-person V_1. The behavior of each was decisively shaped by an uninvited intrusion. Without it, they would not have acted as they did. Since V_3, V_4, and V_5 would not have existed at all, but for D/n's mischief-making, we cannot say how they would have acted in its absence.

A disparity which seems insignificant to me is that the street-crosser knows of the forces constraining him. An important consequence of this knowledge is that our street-crosser makes up his mind, grudgingly, to go along with the wind or his abductors. But V_1, V_2, and V_3, when reacting to D/n's antics or D/n's program, cannot consent to an intrusion which they ignore.

So much for parallels with straightforward compulsion. If we look back at our original case of doing what one wants, and forward to instances of disturbed motivation, where do our transitional anecdotes belong? Neither in the plain example of wanting and doing, nor in the most neurotic cases, do we spot an interference – of anything like hurricanes, coercers, hypnotists, or D/n's. Possibly the cravings and 'need' symptoms of some addicts and alcoholics result inevitably and principally from the absorption of enough drug or drink. They may become addicted by taking medicine which they do not know to contain narcotics. This happened to Coleridge. Some wayward impulses can be traced to a vitamin deficiency, to low blood sugar, to a definite trauma or Oedipal situation. As with V_3, V_4, and V_5, we cannot guess how the motivationally disturbed person would have behaved on this precise occasion, if only his parents had not done such-and-such 28 years ago. Hence our bridging cases differ both from ordinary instances of doing what one wants, and from paradigms of acting upon a wayward desire. Accordingly, I doubt that we shall be able to analyze most motivational disturbances as similar to hypnotism, D/n-ological monkeying, or garden-variety compulsion.

But we should see if we can single out infra-psychic counterparts of the external interventions which we have reviewed, and to explain how these might overwhelm a person's 'real' attitudes.

VIII. TYPES OF MOTIVATIONAL UPSET

I shall not summarize even the American Psychiatric Association's catalogue of 'standard' diagnostic labels – from 'brain disorders,' 'psychotic disorders,' 'personality disorders,' to 'mental deficiencies.' I shall use the term 'attitude' to cover all conative, affective, and cognitive states on which people may be said to act. What interests me are attitudes of the person which strike him, in Freud's words, as 'from an alien world' – either while he is acting on them, or retrospectively. I have followed tradition and concentrated on bizarre impulses and urges. No doubt this is the addict's and kleptomaniac's burden. But peculiar beliefs, even beliefs about who one is, often dominate causes where we feel that the person has lost command of his behavior. Take a paranoid. Given his strange but firm conviction that people are eavesdropping on him, conspiring against him, poisoning him, maybe directly controlling his thoughts and movements, aren't his self-protective desires appropriate enough? Or consider people who slip into a 'fugue' state, hallucinate all sorts of extreme dangers, and 'fight back,' thereby sometimes unknowingly harming innocent bystanders. Aren't such aggressive impulses correct for the imagined threatening circumstances? Only these people's beliefs, derived from wildly distorted perception, seem out of whack. Again, a woman afflicted with 'multiple personalities' will 'black out.' During longer or shorter periods of time she will forget who she is, and believe that she is some other, actual or imaginary, woman or man. When she 'comes to,' and once more has accurate beliefs about her identity, she will have no beliefs at all – no memory – regarding what she did and thought during the 'blank' interval. Her desires during the interval may have been innocuous, and suitable to whatever she believed.

We mentioned paranoid individuals who believe that their thinking is controlled by others. One patient during the 1950s imagined that

> she has been receiving telepathic messages . . . This experience is denominated, in her language, 'airings.' . . . She has received 'airings' from, among other people, President Eisenhower . . . . Thoughts were poured on her by some machine done by her enemies ([10], pp. 431ff).

A Freudian example of someone being troubled by thoughts which strike him as not his would be 'the familiar case of sacrilegious thoughts entering the minds of devout persons' ([9], 1909d, X, pp. 193, 242f).

IX. DO PSYCHOANALYTICAL THEORIES HELP US SINGLE OUT 'MY' MOTIVES?

Although I have drawn from Freud, I cannot digress to evaluate his challenging hypothesis that most psychotics, neurotics, and 'normals' totally ignore those fantasy-thoughts, fears, and urges which most profoundly shape their conduct. Speaking roughly, these 'repressed' items make us do, also usually unawares, a vast number of things; or else make us devise superficially plausible rationalizations for consciously doing what we have a repressed need to do. Freud's innovative concept of action will have to be neglected as well. But I should note that it seems to encompass nearly everything that common sense, jurisprudence, and philosophy contrast with our deeds. Earlier I assumed the polarity of acting and being acted upon – for example, being carried. However, Freud suggests that many misfortunes which other people seem to be inflicting upon us should actually count as things we do. Some people have an unconscious 'compulsion to repeat' childhood fantasies and traumas:

> their fate is for the most part arranged by themselves and determined by earlier infantile influences .... Thus we have come across people all of whose human relationships have the same outcome: such as the benefactor who is abandoned in anger after a time by each of his protégés ([9], 1930a, XXI, p. 117).

The same is true of blunders and 'breakdowns' which may spoil our purposive endeavors, and which we presumably intend to avoid: slips of the tongue while speaking; the botching of tasks we are trying to complete; misremembering; accidentally injuring ourselves. We seldom regard these mishaps as actions in their own right, alongside the performance they mar. But Freud reclassifies them as either 'unconsciously intended' deeds, or as acts of 'compromise' between our warring conscious and repressed purposes. And of course where we see the physical 'symptoms' of neurosis as mere bodily processes, Freud re-baptizes these tics, tremors, psychogenic paralyses and pains as things we do – albeit unconsciously.[3]

While neglecting all this, we can test the psychoanalytic approach to our cases of people who knowingly perform 'ordinary' deeds because they are afflicted by conscious attitudes which seem foreign to them. But I suspect that psychoanalytic theories of the self cannot possibly separate our own motives from wayward, unrepresentative drives which overpower us.

For psychoanalysis, the individual is an organized collection of forces and either mechanical or anthropomorphic components which dynamically interact. No element is more representative of him than any other. To illustrate: Freud seems to see tension as the primary feature of mental life: 'Only by the concurrent or opposing action of the two primal instincts the erotic and the destructive . . . can the motley variety of vital phenomena be explained' ([9], 1937c, XXIII, p. 243). Freud's early work presents a tug-of-war between the 'pleasure principle,' or 'pleasure ego', which drives us toward immediate satisfaction, and the 'reality principle,' or 'reality ego', which aims 'for what is *useful*' ([9], 1911b, XII, p. 223). Other tensions build up among parts of our psychic apparatus: the conscious, preconscious, and unconscious 'systems' of ideas, in Freud's first doctrines; and the quite differently structured ego, superego, and id, after 1923. Initially Freud casts our 'ethical and other standards' as 'repressing forces' to keep our unsavory ideas at bay ([9], 1910a, XI, p. 24). In mechanical terms, 'the repressed exercises a continuous pressure in the direction of consciousness, . . . [which] must be balanced by an unceasing counter-pressure' ([9], 1915d, XIV, p. 151). Freud's later theory is dominated by social metaphors. The 'poor ego' of mature tripartite theory 'serves three masters and does what it can to bring their claims and demands into harmony . . . Its three tyrannical masters are the external world, the superego, and the id . . . [I]f it is hard pressed, it reacts by generating anxiety' ([9], 1933a, XXII, p. 77). The upshot will be either neurosis or psychosis, depending upon 'whether . . . the ego remains true . . . to the external world and attempts to silence the id, or whether it lets itself be overcome by the id and thus torn away from reality' ([9], 1924b, XIX, p. 151). So motivational disorder seems to be mainly a sharpening of the internecine struggles which typify mental life.

Now I suppose we tend to equate ourselves with the early 'conscious' system of ideas, and with the ego of mature doctrine. After all, when a thought or impulse passes through my 'consciousness,' I must be aware of it; and I am certainly unaware of ideas belonging to my unconscious system. As for my ego, Freud is careful to explain that it carries out many tasks of which I cannot possibly be conscious – notably repression 'in the service . . . of its superego,' and resistance to recovery from neurosis ([9], 1923b, XIX, p. 52, 49ff). But still, it is called the 'I' (*das Ich* in German); and I am surely grateful for its peace-keeping efforts as well as its devotion to 'reality.'

Freud's remarks about therapy encourage our identification with consciousness or ego. He says the psychoanalyst 'works hand in hand with one part of the pathologically divided personality, against the other partner in the conflict' ([9], 1920a, XVIII, p. 150). The analyst's objective is to 'give . . . back . . . command over the id' to the neurotic's oppressed ego ([9], 1926e, XX, p. 205).

But we cannot assimilate self to consciousness or to ego. For the point of

Freudian personality theories is that we are also made up of unconscious, libidinal, aggressive, puritanical, and other elements. None singly constitute the person. None has any special primacy. Genetically, unconscious instincts and the id take precedence. Thus Freud says that the 'core of our being . . . is formed by the obscure id . . . ' ([9], 1940a, XXII, p. 197). Speaking evaluatively, he allows that the 'bad repressed contents' of my dreams belong 'to an "id" on which my ego is seated;' however, 'this ego developed out of the id, . . . forms with it a single biological unit . . . , [and] obeys . . . the id' ([9], 1925i, XIX, pp. 133f). So psychoanalysis only entitles us to regard the person, his attitudes and behavior, as a smoldering mixture of psychical forces and components. No item within me is the real me. More important, none corresponds to the external forces and coercers that sometimes compel me to do the least of evils, when I would prefer not to be in that kind of choice situation.

X. DESIRES THAT CONFLICT WITH MY VALUES

At least Freudian theories of the self should inhibit philosophers from assuming that attitudes belong to us if they harmonize with our morality, and do not belong to us if they clash with it. Yet the latest account of how desires can exert compulsion, by Professor Gary Watson, seems to deploy this very strategy. Watson holds that

> when actions . . . are unfree, the agent is unable to get what he most wants, or *values*, . . . due to his own 'motivational system' . . . . The strength of one's desire may not properly reflect the degree to which one values its object . . . . It is possible that sometimes [one] is motivated to do things he does not deem worth doing. This possibility is the basis for the principal problem of free action: a person may be obstructed by his own will ([17], pp. 206, 210, 213).

Watson seems aware of the obvious psychoanalytical rejoinder, when he admits that 'one may be as dissociated from the demands of the super-ego as from those of the id' ([17], p. 214). But his worry is that one's moral 'attitude' may have 'its basis solely in acculturation . . . independently of [his] judgment' ([17], p. 214; see [12], p. 42). Surely Freudians would reject Watson's distinction between norms based on conditioning, and norms based upon our own 'judgment.' Watson does not rebut their belief that all our values result from socialization, or from a 'comprise' between it and our instinctual drives. Watson merely compares values deriving from 'acculturation' and those we reason out in a cool and non-self-deceptive moment .... That most people have articulate 'conceptions of the good', coherent life-plans, *systems* of ends ... is of course something of a fiction. Yet we all have more or less long-term aims and normative principles that we are willing to defend .... These ... are ... our values ([17], p. 215).

Now suppose that Watson could prove that 'acculturation' does not significantly warp these ratiocinations. We shall also grant that everyone must have some values. The major psychoanalytical objection remains unanswered: Why assume that our 'life-plans' and '*systems* of ends' are more representative of us than our most savage, incoherent urges? Are not both equally characteristic of us? Perhaps we *value* our disposition toward life-planning more highly than our spontaneous, often self-destructive impulses. But that smacks of circularity. And from the fact that we prize life-planning, even that it is somehow most worthy, we cannot legitimately infer that it corresponds to what we really want. So we cannot say, as Watson does, that we are 'obstructed,' or compelled, and thus unfree, when we act on desires that clash with our values.

XI. ARE 'ALIEN' MOTIVES THOSE WHICH ARE INALTERABLE BY REASONING?

There is a simpler account than Watson's, which omits reference to values. It appears in the essay by MacIntyre cited earlier (Section VII). MacIntyre's thesis seems to be that we act freely when we act rationally:

Behavior is rational – in this arbitrarily defined sense – if, and only if, it can [in principle] be influenced or inhibited by the adducing of some logically relevant consideration ... What is logically relevant will necessarily vary from case to case. If Smith is about to give generously to someone who appears to be in need, the information that this man ... has in fact ample means, will be relevant ... An impulsive action can in this sense be rational ... [And] behavior can be reflective without being ... rational. For a man may spend a great deal of time thinking about what he should do, and yet refuse to entertain a great many logically relevant considerations ([11], p. 248).

MacIntyre neglects to specify whose criteria for 'logically relevant considerations' we should use in establishing an individual's rationality or irrationality. Should we rely on his interlocutor's standards? Should we call in experts? Luckily Professor Neely has quite recently devised a criterion like MacIntyre's for calling some desires 'irresistible,' and others 'more intimately related to the self' ([12], p. 43). The latter do not compel us when we act on them. Neely opts for the agent's own criteria of logical relevance. Taking Socrates' desire to stay in prison and drink the hemlock, Neely admits that his listeners

might well feel that Crito presented Socrates with good and sufficient reasons to escape. Yet ... Socrates' decision to remain ... [is hardly] a clear case of an unfree decision

. . . This leads us . . . [to suggest[: a desire is irresistible if and only if it is the case that if the agent had been presented with what *he took to be* good and sufficient reason for not acting on it, he would have acted on it ([12], p. 47).

Does this yield a satisfactory analysis of motivational disorder? It makes some headway. It explains why 'willing,' 'unwilling,' and 'wanton' drug-addicts, who respectively approve, abhor, and feel indifferent about their habit, all act unfreely. A minor defect might be the subjectivism of accepting whatever the agent '*took to be* good and sufficient,' or 'logically relevant considerations. ' May he not, either consciously or unwittingly, insist upon absurdly high standards of germaneness and conclusiveness? Thus, in MacIntyre's phraseology, he 'refuses' to ponder seriously a vast range of what people generally judge to be sufficient reasons for altering his behavior. He might even agree, later on. But could he have held absurd standards of relevance in a 'cool hour?' The thermal metaphor is unhelpful. People are sometimes coolly obtuse and stubborn. So we will have cases where everyone else thinks the agent is irrationally acting on an irresistible desire, but his denial must prevail.

The principal shortcoming of Neely's criterion for irresistibility of attitudes is one that it shares with the 'values' test. Both criteria – and even an objective, public standard of amenability to persuasion – seem to rest upon a key premise which we attacked above (Section X). If you say that people act freely, and do what they really want to do, when their motives are alterable by reasoning, are you not assuming that the core self is the rational self? For you equate unmodifiable with 'irresistible' attitudes, and imply that these are not really the person's. This entails that impulsive, obstinate, and unpersuadable agents never act freely, never do what they really want. But as we objected above, this questionably rationalistic concept of the person and his genuine attitudes seems to arbitrarily write off his dark, instinctive, irrational side.

XII. HIERARCHICAL MODELS AGAIN

I shall close by reconsidering the double-decker analysis. It came up when we tried to understand the attitudes of someone who is 'externally' compelled to act. Hierarchy theories offered to explicate why a person who submits to a threat, for instance, 'invariably does something which he does not really want to do' ([7], p. 81). Split-level strategists equated 'not really wanting' either with an aversion toward one's basely prudential reason for giving in (Dworkin), or else with a 'second-order volition' to 'have a different motive for acting'

(Frankfurt). What did this accomplish? In effect, for 'he does not really want to' they substituted 'the real he does not want to,' and made one's ground-floor motive the object of his real self's loathing. As an account of people's conative attitudes when they are forced to act, it seemed gratuitous, misleading, and obscure (see Section V). Consequently I proposed a non-hierarchical alternative which retained some of its insights. But perhaps a theory of motivational disturbance, or 'inner compulsion,' has to recognize planes of desire and even 'volition.' After all, in most of the examples of neurosis and psychosis that we reviewed (Section VIII), not to mention cases of 'unwilling' alcoholism and drug addiction, the sufferer's complaint is not only that he does things which he dreads and regrets; it is also that he is afflicted by ghastly thoughts, moods, and hankerings. At least this phenomenological aspect of motivational disarray, and possibly more, may be suited for hierarchical interpretation.

The initial challenge facing level-splitters is a 'regress' problem. Why identify the person with his second-floor attitudes? Why not climb beyond them, and acknowledge somebody's further volition, for instance, to be the sort of person who is – or is not – stirred by a second-order desire to be 'motivated . . . by the desire to concentrate on his work,' or to be 'moved by kindness?' ([6], pp. 10, 17).

Frankfurt seems to anticipate the 'regress' objection. But his attempt to dismiss it baffles me. He considers someone with an 'unresolved conflict' of second-order desires, who therefore cannot reach a second-order volition. This 'prevents him from identifying himself . . . with *any* of his conflicting first-order desires' and 'destroys him as a person;' the second-order impasse

> either tends to . . . keep him from acting at all, or it tends to remove him from his will so that his will operates without his participation . . . . Nothing prevents an individual from obsessively refusing to identify himself with any of his desires until he forms a desire of the next higher order. The tendency to generate such a series . . . also leads toward the destruction of a person ([6], pp. 15f).

What annihilation is Frankfurt warning of? We recall that he declared second-order volitions 'essential to being a person' ([6], p. 10). Do third-story volitionists only risk the semantical destruction of not meeting Frankfurt's requirement for personhood? I am unsure of this – and equally unsure how inaction, or being 'removed from their will,' might undo superstratospheric volitionists.

Rather than speculate on this, we should look at a triad of more definite arguments by Frankfurt against higher-than-second-order capers. Two arguments begin with someone forming a 'decisive' second-order volition:

he has decided that no further questions about his second-order volition, at any higher level, remains to be asked. It is relatively unimportant whether we explain this by saying that this commitment implicitly generates an endless series of confirming desires of higher orders, or . . . that [it] . . . is tantamount to a dissolution of the pointedness [*sic*] of all questions concerning higher orders ([6], pp. 16f).

The 'no further question,' or 'dissolution,' argument sounds evasive. The 'endless series' alternative, on the other hand, seems to repeal Frankfurt's ban on infinite ascent. However that may be, Frankfurt has a third reply to the 'regress' objection. He invokes the obscure notion of someone being 'active with respect to his own desires when he identifies himself with them.' Frankfurt announces:

As for a person's second-order volitions . . . , it is impossible for him to be a passive bystander to them. They *constitute* his activity – i.e., his being active rather than passive – and the question of whether or not he identifies himself with them cannot arise. It makes no sense to ask whether someone identifies himself with his identification of himself, [except to ask] . . . whether his identification is wholehearted . . . ([8], p. 121).

I concur when Frankfurt adds: 'This notion of identification is admittedly a bit mystifying.' I hope I will be excused for doubting that Frankfurt has solved the 'regress' problem.

But suppose he had. He still ought to justify his tacit assumption that our second-floor attitudes, and the street-level desires they endorse, invariably reflect what we really want. Why say that if we act on any contrary desire, it compels us, and we act unfreely? Frankfurt's account of inner compulsion stirs up all the misgivings we felt toward Watson's thesis that a person's values constitute what he 'most wants' to do, and Neely's ' "being open to persuasion" criterion' for separating one's own from 'irresistible' alien desires. Why assume that the 'valuing' self, the 'rational' self, or the second-order self is preeminently you, and that its deliverances are especially yours? Why are the first-order motives which you anathemize from on high not equally yours?

Certainly Frankfurt's upstairs–downstairs model dramatizes the plight of an 'unwilling addict:' the wretch 'identifies himself . . . with one . . . of his conflicting first-order desires . . . makes [it] . . . more truly his own and . . . withdraws himself from the other;' consequently the latter 'force moving him to take the drug is a force other than his own' ([6], p. 13). But we noticed (Section XI) that a ground-floor 'persuadability' analysis suffices here too. Furthermore, when we discussed the attitudes which a hypnotist or a Devil/neurologist might implant in you, we hinted that some cases of addiction and alcoholism might be analogous (Section VII). Imagine that the drug-taker's craving or 'need' feelings have resulted principally from absorbing his anodyne. Perhaps he was unaware that it was a narcotic. Then we have an utterly simple and unhierarchical explanation why this desire is 'other than his own': it originated from cumulative physical intrusions of the drug. Such ancestry is uncharacteristic of motivational disarray.

But for the sake of argument, suppose that we lacked a one-story model of unwilling addiction, and that such addiction is a prime example of being compelled by a desire. We visualize the hapless drug-taker on his second-floor balcony, committing himself to, or identifying himself with, the weaker anti-narcotics 'force' below. Then he surrenders to its irresistible rival. Do we learn why the victorious craving was not his, and why it was undeterred by his disapproval? The presurrender phases of this miniature tragedy do not intelligibly explain how 'identifying' works.

The non-hierarchical analysis I proposed for external compulsion (Section XI) is not generally applicable to motivational disturbances. Only a most reflective addict would regard 'compliance' with his craving as the lesser evil, and say that he prefers to escape his inner-conflict situation altogether. As for a neurotic person, such as an obsessive, it would make no sense to assert that he only wants to enact his bizarre rituals because that seems the least damaging course left open to him by his overpowering desire.

I conclude that we have devised no adequate analysis of how some people's attitudes make them act unfreely. But we have gotten a clearer understanding of the puzzle, of the key contrast between doing what one wants and being constrained, and of the reasons why we cannot model attitudinal derangement upon coercion.[4]

University of Illinois at Chicago Circle

NOTES

[1] On the social nature of these maladjustments, I follow such writers as Laing, Cooper, Scheff, Szasz, Goffman and Brown, in Brown [1]; also Sayers [14].

[2] For my 'compromise' view, see Essay III, with A. B. Levison, in Thalberg [15].

[3] For more thorough discussion and criticism, see [16].

[4] I read penultimate drafts of this paper at Universtiy of North Carolina, Greensboro, as well as University of Colorado. I thank listeners for criticisms and suggestions, some of which I have incorporated. I also enjoyed exchanges with Professor Caroline Whitbeck, before and during the Galveston conference. In response to her comments, I have thought it fair only to make my own position more explicit than it was in the version she discussed at Galveston. I have not attempted to parry her objections. But from my skeptical approach to the assumption that a person may be forced by his own motives to do something he doesn't really want to do, it is obvious that I would challenge

any unexplained assumption of Professor Whitbeck's that a person may be deceived by himself into believing something he doesn't really believe.

BIBLIOGRAPHY

1. Brown, P. (ed.): 1972, *Radical Psychology*, Doubleday, Garden City, New York.
2. Dworkin, G.: 1970, 'Acting Freely', *Nous* **4**, 367–383.
3. Dworkin, G.: 1976, 'Autonomy and Behavior Control', *Hastings Center Report* **8**, pp. 23–28.
4. Feinberg, J.: 1970, *Doing and Deserving*, Princeton University Press, Princeton, New Jersey.
5. Flew, A.: 1955, 'Divine Omnipotence and Human Freedom', in Flew *et al.* (eds.), *New Essays in Philosophical Theology*, SCM Press, London, pp. 144–169.
6. Frankfurt, H.: 1971, 'Freedom of the Will and the Concept of a Person', *Journal of Philosophy* **68**, 5–20.
7. Frankfurt, H.: 1972, 'Coercion and Moral Responsibility', in T. Honderich (ed.), *Essays on Freedom of Action*, Routledge, London, pp. 72–85.
8. Frankfurt, H. and Locke, D.: 1975, Symposium: 'Three Concepts of Free Action', *Proceedings of the Aristotelian Society*, Supp. Vol. 49, 95–125.
9. Freud, S.: various dates, in J. Strachey (ed.), *Standard Edition of the Complete Psychological Works of Sigmund Freud*, Hogarth Press, London, 1954–74, 24 vols.
10. Katz, J., *et al.* (eds.): 1967, *Psychoanalysis, Psychiatry and Law*, Free Press, New York.
11. MacIntyre, A.: 1957, 'Determinism', reprinted from *Mind*, in B. Berofsky (ed.), *Freedom and Determinism*, Harper and Row, New York, 1966, pp. 240–254.
12. Neely, W.: 1974, 'Freedom and Desire', *Philosophical Review* **83**, 32–54.
13. Nozick, R.: 1969, 'Coercion', in S. Morgenbesser *et al.* (eds.), *Philosophy, Science and Method*, St. Martin's, New York, pp. 440–472.
14. Sayers, S.: 1973, 'The Concept of Mental Illness', *Radical Philosophy* **5**, 2–8.
15. Thalberg, I.: 1972, *Enigmas of Agency*, Allen and Unwin, London.
16. Thalberg, I.: 1974, 'Freud's Anatomies of the Self', in R. Wollheim (ed.), *Freud*, Doubleday, Garden City, New York.
17. Watson, G.: 1975, 'Free Agency, *Journal of Philosophy* **72**, 205–220.

CAROLINE WHITBECK

TOWARDS AN UNDERSTANDING OF MOTIVATIONAL DISTURBANCE AND FREEDOM OF ACTION: COMMENTS ON 'MOTIVATIONAL DISTURBANCES AND FREE WILL'[1]

In his paper, Professor Thalberg focuses upon one sort of circumstance under which we might wish to speak of motivational disturbance, namely, that in which a person has urges, feelings, and perhaps even beliefs which strike that person as foreign. Thalberg uses the term 'attitude' to apply to whatever urges, feelings, etc., we may be said to act on. I take it that these are said to be 'foreign' or 'alien' because, although perhaps they are *familiar*, they seem to the person to be incoherent with the rest of the person's beliefs, feelings, and (I would add) habits. The question regarding motivational disturbances which interests Thalberg is how such attitudes make the actions which they engender unfree, or, at any rate, less free. The aspects of this question which Thalberg addresses in this paper is whether the constraint in these cases is very much like that which exists in cases of outright coercion. There is a prima facie resemblance between cases of motivational disturbance as defined above, and cases of acting under coercion in that in both cases the person performs the action, as opposed to merely being acted upon. But in each sort of case an alien factor is 'crucial' in Thalberg's sense, i.e., it is necessary if the act is to be performed at all. In each sort of case we might want to regard this as a constraint which interferes with acting freely, i.e., as something which interferes with doing what one wants. Thalberg thinks that we can. However, as Thalberg points out, it is rather paradoxical to say that people are constrained in cases of motivational disturbance, since in such cases the person wants (or has an impulse, etc.) to do something and does it. But *wanting to do something and doing it* is usually what is thought to characterize acting freely. Thalberg then offers a non-hierarchical analysis of coerced action, and argues that neither in the case of coerced action, nor in the case of action due to disturbed attitudes, is it correct to say that the person really did not want to perform the action in question. However, he argues that there the analogy ends, and that motivational disturbances are not amenable to the same sort of analysis as coerced action. The issues raised in Professor Thalberg's paper may help us better understand, first, how free action is compromised by motivational disturbance, and, second, how diminution of these conditions increases freedom of action. I shall be somewhat selective about the parts of his paper on which I comment upon in detail.

H. T. Engelhardt, Jr. and S. F. Spicker (eds.), Mental Health: Philosophical Perspectives, 221–231.

It is disappointing that Thalberg does not really say very much about motivational disturbance, and what he does say is either not entirely clear, or else does not apply to some of the phenomena which he purports to analyze. We are offered arguments to the effect that certain hierarchical analyses of *coerced action* are less economical than his own non-hierarchical analysis, but we have few clues as to how we *should* understand *motivational disturbance*.

I will begin by considering his examples of alcoholism and drug addiction and argue that these examples do not answer to his description of motivational disturbance. Self-deception plays a role in alcoholism and drug addictions and such self-deception may compromise free action, but Thalberg's definition of motivational disturbance readily applies only to cases of so-called neurosis. Having said that, I should make explicit that, unlike Thalberg, I do not assume that there are *no* so-called mental illnesses which should be understood as disease entities.[2] It seems to me that there is no case at all for regarding the so-called personality disorders as disease entities, that the case for regarding neuroses in this way is a very weak one, and that psychoses *are* diseases or manifestations of disease. (The term 'psychosis' is ambiguous, meaning either a type of syndrome or a type of disease entity.) However, in the case of psychoses it is not clear which of these are *mental*. This decision is based upon etiology, and etiology is often established only for organic psychoses (i.e., such as toxic psychoses and psychoses due to brain tumors). It is not clear whether psychosocial factors play such an important role in the etiology of any of the psychoses that we will want to regard *these* as constituting the etiologic agent, and accordingly classify the psychosis as a *mental* disease. I have argued elsewhere that medical thinking is moving away from the expectation that there exists a unique etiologic agent definitive for each disease entity.[3] If this is so, then what are now regarded as etiologic agents will come to be regarded as contributory causes, and not definitive for the disease. In that case there will be no basis for maintaining the distinction between mental diseases (diseases whose etiologic agents are psychosocial factors) and diseases in which psychosomatic factors play a role. (My view will no doubt be unintelligible to Cartesian dualists who postulate a dichotomy between the mental and the physical.)

To better understand what phenomena do answer to Thalberg's description of motivational disturbance, consider the case of a person who has a congenitally high arousal level which manifests itself in a generally high anxiety level, born into a family many of whose members have a similar disposition to be anxious. Further, suppose that these family members have maladaptive ways of coping with this anxiety, e.g., smoking, overeating, yelling at one another, etc. Further suppose that, having picked up some of these habitual responses, the person comes to realize how these further debilitate her or him. I take it that the inclination to yell, smoke, or overeat would still arise in the person in question. But the person could try to replace the maladaptive responses with more adaptive ones. It seems to me that it is a mistake to say that such a person has a motivational disturbance in the sense defined, i.e., that the person has experience which s/he regards as alien. On the contrary, it is perfectly clear to the person that the inclinations in question are to do things which would achieve the desired end, that of relieving anxiety by releasing tension or through sedation. However, since the means to these ends would have destructive effects on the person's health or social relations, that person would like to replace the habitual resort to these means with the habitual use of other means.

I certainly would not deny that people may often fail or only partially succeed in reforming their habits, but such failure counts as weakness of the will and not as motivational disturbance. And although weakness of the will is an interesting philosophical topic, it is not the one we have before us today.

Alchoholism and drug addiction are problems of the same sort as that presented in the above example, only the alternative means to satisfaction seem harder to come by. Typically what addicts and alcoholics want is to feel successful, secure, powerful, attractive, or loved. And actually becoming successful, etc., and accepting that fact on the affective or feeling level is no mean feat. Adding the issue of physiological addiction really does not change the problem. Physiological addiction, like a high arousal level, creates discomforts but does not itself impel a person to take a certain course of action. Withdrawal may be a painful shattering experience, but then so is the experience of undergoing certain diagnostic tests. It is certainly unfortunate that some people have only the very unpleasant choice of undergoing a painful experience or continuing a course of action which may lead to degeneration or death. A person faced with such limited options is for that reason less free than one with more options, but this is not because the person's actions are influenced by alien feelings, impulses, etc. In general, Thalberg's definition of motivational disturbances makes it applicable to what are called 'cases of neurosis' but *not* to cases of what are called 'personality disorders.' You may recall that one of the principal features which distinguish the so-called personality disorders from the neuroses is that the feelings, impulses, behavior, etc., that are labeled as symptoms are regarded by the person as compatible with the rest of her or his feelings, beliefs, habits, etc. In one set of jargon they are said to be 'ego-syntonic' as opposed to 'ego-alien.'

Self-deception may play a particularly important role in such so-called personality disorders as addiction, and since I think self-deception interferes with acting freely in many of the types of mental illness that Thalberg mentions, I shall return to this subject below. However, it is worth noting that not *all* so-called personality disorders are marked by self-deception. For example, in our society, any person who is a hobo is considered to have a personality disorder, but, even if one accepts a negative characterization of this life style as one which shows an inability to take responsibility, establish intimacy, etc., it does not seem to be one that requires self-deception. Therefore, even if we expand our concern and examine the way in which self-deception compromises our action, as well as the way in which motivational disturbance does so, we will not have shown any way in which those with so-called personality disorders should be thought to act less freely than others. (I take it that this finding supports the view that it is peculiar even to call such personality types 'illnesses,' much less 'diseases'.)

What about psychoses? Do these involve anything that answers to Thalberg's characterization of motivational disturbance? In cases of psychosis it is not clear that there is a relatively stable well-defined conscious self, so that the question of whether certain impulses, etc., are congruent with such a self loses its meaning. (The view that in psychosis there is no such self is prevalent in certain contemporary psychoanalytically oriented psychiatry. Rather than postulating that there is a well-defined ego which is overwhelmed by, or which, as Freud put it, 'throws itself into the arms of unconscious impulses' [2] p. 27, many contemporary theorists say such things as 'the ego is not well-defined,' or 'the ego boundaries break down' in psychosis.)

In some cases a person experiences fairly well-defined psychotic episodes, and between episodes seems as normal as anyone. In the case of *such* people one might wish to say that during their non-psychotic period they are themselves, and relative to those selves their so-called psychotic impulses, feelings, and even *beliefs* are regarded as alien, much as we imagine Ulysses *in retrospect* regarded as alien the urges, feelings, etc., which he experienced on hearing the Siren song. Now in such cases, where there is no evidence of psychosis between episodes, we have reason to suspect that the psychoses may be due to toxic substances (i.e., acute organic brain syndrome). To get as clear an example as possible, let us consider a case in which some substance has an effect on a particular person which is similar to that which follows ingestion of LSD in the general population. Now the problem is that the periodic LSD-type reaction seems to dissolve psychic boundaries so that during the psychotic episodes the person seems less like Ulysses, who supposedly experienced overwhelming impulses, etc., and more like the person who becomes delirious as a result of a fever. In cases of delirium it is not clear that one can say that the person *acts* any more than that someone acts when s/he is undergoing a grand mal seizure. Therefore I conclude that the case for assuming that *any* psychoses can be construed as cases of motivational disturbance is not a clear one.

If we decide against including psychoses among the types of motivational disturbances I would argue that we will have to strike 'beliefs' (and therefore 'delusions') although not 'thoughts' from the list of mental entities which we might both *have, and regard as foreign*. Beliefs, unlike thoughts, cannot be entertained without commitment. Beliefs are propositions to which one assents. Therefore, to simultaneously *have* a belief *and* to regard it as alien is a contradiction. The only way a belief or delusion can be alien is for it to be believed at one time and regarded as alien at another. Thus the person subject to psychotic episodes might regard the beliefs held during a psychotic episode as alien, and *if* we could regard the intrusion of the whole psychotic state of mind as motivational upset, we could speak of the intrusion of alien beliefs.

In summary, I do not consider alcoholism or drug addiction (or *any* of the other personality disorders) to be motivational disturbance as Thalberg generally does, nor do I think that the case with respect to psychosis is a clear one. Of the primary types of mental illness only the neuroses answer to Thalberg's definition of motivational disturbance. I shall return to the subject of neuroses below, but first I wish to consider the comparison between coerced action and action motivated by alien impulses, urges, and feelings.

I concur with Professor Thalberg in believing that the particular hierarchical accounts offered by Frankfurt and Dworkin have defects. However, I think that we may have to make reference to volitions of second (and perhaps higher) order to adequately understand the extent to which a person acts freely, and that Thalberg has insufficiently appreciated the plausibility of hierarchical accounts in general. This failure is due in part to the way he has framed some of his examples. Thalberg asks, 'What is the holdup victim more opposed to: being deprived of his money, or abandoning it "for these reasons"?' and replies, 'Surely the former! Again, is it not less painful to act "for these reasons" than to lose one's billfold?' ([4], p. 205). In describing the holdup victim as having been *deprived* of his money, Thalberg has covertly made reference to his reasons for giving up the money. One would want more of an account of the holdup victim's circumstances to be sure of just what would bother the person most, but I think it likely that for most people, we can imagine circumstances in which they would gladly hand over their billfolds, say to do a friend a great benefit.[4]

When we consider the bearing of the hierarchical analysis upon the general issue of freedom of action, we find that the definition of acting freely as acting for reasons one does not mind acting from, which Thalberg quotes from Dworkin, fails in an interesting and important way to capture what we mean by acting freely. This failure shows how necessary *some* sort of hierarchical analysis is for an understanding of at least some human actions. Dworkin's definition would allow that a person who changes goals so as to have acceptable reasons for acting in a given way, and acts for these reasons, would be said to act freely. But that is not necessarily the case as is illustrated by the speech from *Iphigenia in Aulis* quoted below. You may recall that in Euripides' play Agamemnon intends to sacrifice his daughter Iphigenia to the goddess Artemis so that he might successfully sail against Troy. Achilles, upon hearing of this, offers to defend Iphigenia against her father. In this speech Iphigenia tells her mother that she wishes to refuse his offer and asserts that she *freely* chooses to go to her death.

> Mother, let me speak!
> This anger with my father is in vain,
> Vain to use force for what we cannot win.
> Thank our brave friend for all his generous zeal,
> But never let us broil him with the host,
> No gain to us, ruin for himself.
> I have been thinking, mother, –hear me now!–
> I have chosen death: it is my own free choice.
> I have put cowardice away from me.
> Honour is mine now. O!mother, say I am right.
> Our country – think, our Hellas – looks to me,
> On me the fleet hangs now, the doom of Troy,
> Our woman's honour all the years to come.
> My death will save them, and my name be blest,
> She who freed Hellas! Life is not so sweet.
> I should be craven. You who bore your child,
> It was for Greece you bore her, not yourself.
> Think! Thousands of our soldiers stand to arms,
> Ten thousand man the ships, and all on fire
> To serve their outraged country, die for Greece:
> And is my one poor life to hinder all?
> Could we defend that? Could we call it just?
> And, mother, think! How could we let our friend
> Die for a woman, fighting all his folk?
> A thousand women are not worth one man!
> The goddess needs my blood: Can I refuse?
> No: Take, it, conquer Troy! This shall be
> My husband, and my children, and my fame.
> Victory, mother, victory for the Greeks!
> The foreigner must never rule this land.
> Our own land! They are slaves and we are free.

([1], 1, 1368–1400)

We must recognize Euripides' ironic purpose in much of this speech, but nonetheless I think we can see much that is typically human in Iphigenia's reasoning and behavior. She rejects Achilles' offer, believing that he could not save her anyway, and maintains that she *freely chooses* to sacrifice herself, for by doing so she will secure benefits to Greece, and thus honor for herself. Let us suppose that Iphigenia is not merely rationalizing her decision to submit to sacrifice but is sincere in her assertion that the goals of victory for Greece (and perhaps honor for herself) are *now* paramount. Still I do not think we would say that she is acting freely in sacrificing herself, for we believe that her decision to embrace these new goals was not a free decision. Because Iphigenia does 'care about her will,' as the hierarchical theorists claim, she wants neither to give herself over out of despair, nor presumably to be dragged willy nilly to slaughter. Coercive circumstances then force her to make an overriding investment in goals which we believe *she would not otherwise have adopted*. The case is not one of motivational disturbance, since Iphigenia did not experience the desire to sacrifice herself to procure victory until she had identified it as her own. However, Iphigenia is still acting under coercion, for although she meets Dworkin's criteria for acting freely, that is, she is acting from reasons that she does not mind acting from, her investment in the goal of securing victory for the Greeks was itself coerced.

Iphigenia is at least mistaken about the freedom with which she 'chooses' to be sacrificed. (Insofar as she *chooses* to ignore the fact that her adoption of the goal in question was coerced, she deceives herself.) She also deceives herself about her external circumstances: she has no reason to believe that the Greeks were themselves threatened with enslavement by the Trojans. The latter sort of self-deception is extraneous to my purposes. It only draws our attention to the desperation with which Iphigenia seeks to find acceptable reasons for sacrificing herself. Had she found an interest in Greece's victory an acceptable reason by itself without inventing the threat of Greek enslavement, her speech still would have served to show that Dworkin's definition fails to capture what we mean by acting freely.

I think that the Iphigenia example illustrates three important points. First, it shows us that *some* sort of hierarchical analysis will be necessary to account for certain aspects of human actions, since Iphigenia's actions are intelligible only on the assumption that she *does* 'care about her will' – it is not just that people do sometimes care about their will, but that caring about it is intimately related to counting oneself as free (but not to being free). Second, it shows that while people often *do* desire to act for certain reasons and not others, and *count themselves free* when they so act, they may not in fact be free even when they are *sincere* in their claims to believe that these acceptable reasons are good reasons and *their* reasons for performing the act in question. We must know more about the process by which they came to accept the goals in question as their own if we are to determine the extent to which *these goals* were freely adopted. Only then can we assess the extent to which the person acts freely in a given situation in acting for these (acceptable) reasons. Finally, since automony and the absence of self-deception are dimensions of mental health, the second point shows us, not only that a hierarchical analysis is likely to be necessary for understanding certain types of mental illness or problems in living, but also that our tendency to represent ourselves as free *may itself lead us to mistake the extent of our own freedom* (and perhaps deceive ourselves about external circumstances), and, therefore, to make us less free. To put this third point another way: the example of Iphigenia shows us that while in many circumstances the assumption that we freely choose to do what we do may lead us to take more responsibility for our lives and act more autonomously, there are other circumstances in which the tendency to represent ourselves as free agents leads us to misunderstand the real choices we have. This tendency has especially disastrous consequences for those who, like Iphigenia, are oppressed and exploited and who, rather than recognize the extent to which coercive factors operate in determining their actions, come to affirm that they choose to be exploited.

Because *some* sort of hierarchical analysis will be necessary to account for the phenomena discussed in the second point above, it seems to me that a non-hierarchical analysis such as that proposed by Professor Thalberg for cases of coercion cannot give a good account of cases like that of Iphigenia. Furthermore, his analysis of acting under coercion seems to suffer from a subjectivism similar to that which infects Dworkin's definition of acting freely. Consider a hypothetical example of two identical twins reared separately. Suppose that the first has lived her whole life in circumstances which give her only very limited options. To bring the case closer to actual mental health concerns, suppose her opportunities for human relationships include none that are mutually supportive. Suppose that the second grew up in exactly the same sort of circumstances, but for a brief period lived in circumstances which afford her the opportunity to form mutually supportive relationships. Further suppose that the first has read about people living under such conditions, so she has some idea of what it would be like to have such options. Finally, suppose that the second twin again finds herself in the old circumstances. We may assume that the second twin would have a keen sense that she does not 'really want' any of the sorts of relationships open to her, but would rather be out of these circumstances altogether. On the other hand, the first twin, while she might day dream a bit about having other sorts of relationships, would not think to list being out of her circumstances altogether as her top preference when considering what relationships to pursue. Thalberg's analysis would lead us to say that the first acts freely but that the second is coerced in choosing among the impoverished relationship open to her, and this seems just wrong. *If either* acts more freely in these circumstances, I would think it would be the second twin, the one with the greater store of experience.[5]

Finally, what should be said about the usefulness of psychoanalytic models in understanding motivational disturbances? Given my earlier argument to the effect that Thalberg's definition of motivational disturbance applies to neuroses, rather than psychoses or personality disorders, and given that it is with neuroses that psychoanalytically oriented therapies have their greatest success, it seems reasonable to look to psychoanalytic theory for the rudiments of an account. I am convinced by the arguments which Thalberg has presented elsewhere that most of what Freud has said about the relation between the ego, superego, and id is incoherent [3]. However, even if we reject this tripartite model of the self, and with it the view that one of these personified structures might do something like coerce another in cases of motivational disturbance, still there is some point to retaining the view that in addition to the conscious and more or less rational self, there may be a good deal of cognitive, conative, and affective material which is unconscious in virtue of having been repressed. (I leave it open here whether there may be additional unconscious material.) The representation at the conscious and behavioral level of this repressed material is symbolic, and it is the symbolic character of this material which is its hallmark and which makes it seem alien to the conscious self. (I would grant that in many actual cases of motivational disturbance it is not clear to what extent, the experiences and behaviors which a person disowns *are* symbolic, and thus whether the person's problem is best understood as neurosis or self-deception.) Because of the symbolic character of the 'alien' attitudes experienced in cases of motivational upset, we now have given sense to the assertion that the person, the whole person and not just the conscious part, *really* wanted to do the sort of thing *symbolized* by the symbolic act and not the act at all.[6]

Although on this model there is more to the self than its conscious and more or less rational part, (that is, the part which very roughly corresponds to the Freudian ego) only such a conscious and fairly rational part could be said to *make choices* in the sense in which people make choices. Therefore, assuming that there is such a part which is fairly well defined, and assuming that we are right in thinking ourselves capable of acting more or less freely, then it must be this conscious and fairly rational part that so acts.

In order to decide to what extent people act freely in doing all of the various sorts of things that they do, we would need a fully developed theory of the person and of human action. We are far from having such an account, but I think that in his discussion of the way in which motivational disturbance may compromise the capacity to act freely, Professor Thalberg has helped us to understand some of the complex problems which must be faced in order to frame such an account, if it is to be adequate.

State University of New York at Albany,
Albany, New York

NOTES

[1] I am indebted to Sandra Harding for some very helpful criticisms of the version of this paper which I read at the conference.

[2] Thalberg alludes to his views on this point in footnote 1 of his paper. He has clarified his views on this and several other points in private correspondence prior to the symposium.

[3] I have argued this in 'Causation in Medicine: the Disease Model,' delivered to the Boston Colloquium for the Philosophy of Science, April 27, 1976.

[4] It is possible to say that handing over one's billfold in different sorts of circumstances and from different sorts of motives may constitute different sorts of acts, namely, being robbed and benefiting a friend. But since reference to the differences in circumstances and motives is built into these more comprehensive descriptions of the act, this can hardly be used as a reason for thinking that it is *not the reason for acting* which makes the action unacceptable. To put the matter another way: if we refuse to grant that the same act may be done for different reasons, then we render incomprehensible the thesis that it is possible for one's action to be unacceptable because of the reason for acting, rather than the nature of the act.

[5] It may be clear from both the Iphigenia example and the twin example why I have chosen to follow Thalberg's initial formulation of the problem as one concerning the degree to which a person acts freely or autonomously. I think both of these cases make clear that analyses which propose a dichotomy between acting freely and acting under constraint must fail to do justice to the facts.

[6] In the above respect, motivational disturbance *is* like coercion. You may recall that Thalberg claims that if we believe that in our circumstances it is going to be impossible for us to do something we almost never say we want to do it. However, this is a mistake, since we do not take someone to be speaking imprecisely if s/he asserts that s/he wants or 'really wants' a cigarette, but there are none in the house and the stores are closed so s/he will have to do without. Thalberg makes his claim about our use of the verb 'want' in order to pave the way for his non-hierarchical account of coercion according to which we withdraw the common sense claim that the coerced individual desires not to do the act s/he is forced to perform. I have argued that his analysis fails because it makes freedom turn on a lack of appreciation of alternatives. Therefore nothing stands in the way of reinstating the commonsense claim that a person may really want *not* to do something

s/he is forced to do. A person may of course *intend* to do the sort of thing s/he does not want to do, because s/he *does* want to *avoid* other consequences. Such a person may be expected to have a negative attitude toward acting merely to avoid negative consequences, as the hierarchical theorists claim.

BIBLIOGRAPHY

1. Euripides, *Iphigenia in Aulis* 1, 1368–1400.
2. Freud, S.: 1933, *New Introductory Lectures on Psycho-Analysis*, trans. by W. J. H. Sprott, W. W. Norton, New York.
3. Thalberg, I.: 1974, 'Freud's Anatomies of the Self', in R. Wallheim (ed.), *Freud*, Doubleday, Garden City, New York.
4. Thalberg, I.: 1977, 'Motivational Disturbances and Free Will', in this volume, pp. 201–220.

SECTION V

THE MYTH OF MENTAL ILLNESS: A FURTHER EXAMINATION

THOMAS S. SZASZ

THE CONCEPT OF MENTAL ILLNESS: EXPLANATION OR JUSTIFICATION?

> Notable enough too, here as elsewhere, wilt thou find the potency of Names; which indeed are but one kind of such custom-woven, wonder-hiding Garments. Witchcraft, and all manner of Specterwork, and Demonology, we have now renamed Madness, and Diseases of the Nerves. Seldom reflecting that still the new question comes upon us: What is Madness, What are Nerves?
>
> *Thomas Carlyle* (1795–1881)
>
> *Sartor Resartus* ([2], p. 280)

I

Why does the concept of 'mental illness' cause continuing difficulties, both philosophical and practical? There are, as I have tried to show for the past twenty years, several reasons for this [7, 11, 12, 14].

One reason is that 'mental illness' is a literalized metaphor; that is, although minds can be sick only in the sense in which remarks can be cutting, people treat mental diseases much as if they were tiying to carve their steaks with cutting remarks [10].

Another reason is that although 'mental illness' names a role, it is used as if it named a condition; that is, it points to being a patient, but is used as if it pointed to being sick [7, 8].

A third reason is that although 'mental illness' is a prescriptive term, it is usually used as if it were a descriptive one; that is, its actual linguistic function is like that of the phrase 'Please close the door,' but it is widely used as if it were like that of the phrase 'The door is closed' ([8], pp. 49–67).

I have remarked on all three aspects of the problem of 'mental illness,' but have perhaps written more extensively about the first and second aspects of it than about the third. To be sure, these interpretations or misinterpretations of the term are, in actual usage, often combined – for example, literalized metaphor being used strategically as prescriptive-dispositional injunction. Nevertheless, it has seemed to me that I might best fulfill the task assigned to me for this occasion by concentrating on the third aspect of the problem of

H. T. Engelhardt, Jr. and S. F. Spicker (eds.), Mental Health: Philosophical Perspectives, 235–250.

mental illness – namely, on its use as prescription concealed as description, as justification disguised as explanation.

II

Typically, an explanation refers to an event, whereas a justification refers to an act. The difference between these terms is much the same as that between things and persons.

For example, we might ask 'How did lightning kill Jones?' We might then be told that it did so by causing him to have ventricular fibrillation and cerebral anoxia.

We might also ask 'Why did lightning kill Jones?' We might then be told that it was because he continued to play golf during a thunderstorm instead of going back to the clubhouse for a drink, as did his friend Smith. It is important to keep in mind that this sort of statement is an assertion about Jones, the victim, not about lightning or some other aspect of the 'cause' of his death.

Our question about why lightning killed Jones may, however, elicit another type of reply, and it is essential that we consider it also. If our interlocutor is a devoutly religious person, or a very mystical one, he might tell us that lightning killed Jones because it was 'God's will' (or something of that sort). What is important about this answer is that it purports to 'explain' an event by assimilating it to the model we use for explaining an action. By imagining God as some sort of superman, death caused by lightning is pictured as God 'taking' a life. This sort of account pleases and satisfies many people because it fulfills the deeply felt human need for legitimizing, or illegitimizing, not only those things that people do to one another but also those that happen to them.

Suppose, however, that Jones was killed not by lightning but by Smith. We might then reasonably ask both how and why Smith acted as he did. The 'how' question seeks to elicit an explanation of Smith's method for causing Jones' death – for example, did he shoot him, poison him, or stab him? What, then, does the 'why' question seek? The usual answer is that it seeks an account of Smith's motives or reasons for killing Jones. But this, as I shall now show, is only partly true. Actually, in asking this sort of question about Smith, people usually want to know several things, among which the most obvious and important are: (1) Smith's avowed aim or reason for his act; (2) his 'real' reason; (3) the authorities' official account of the reason; (4) the psychiatrist's expert opinion about the reason; (5) the defense attorney's claim about the reason; and (6) the jury's judgment about the reason.

Each of the above reasons is, strictly speaking, a claim or a conjecture; none is an explanation or a cause, in the sense in which these latter terms are understood and used in natural science. Nevertheless, when confronted with this sort of situation, most people feel, as it were 'instinctively,' that one or another of the reasons listed is 'true,' and that the others are 'false.' In fact, they may all be 'true,' in the sense that each represents the sincere conviction of the speaker; or they may all be false, in the sense that Smith acted for reasons, perhaps known only to himself, other than any of those articulated in the several conjectures. Let me illustrate this with a simple example.

Suppose that a person observing patrons ordering food in a restaurant is asked why one of the customers, named Smith, ordered hamburger rather than lobster. The observer would, of course, first ask Smith, who might explain that he did so because he prefers hamburger to lobster. The observer himself might conjecture that it was because hamburger is cheaper. Who really knows why Smith chose as he did? In the sense in which we can know the chemical composition of hamburger or lobster, no one can know why anyone orders one or the other. The only honest answer to this sort of 'why' question is to give an account of the reason for the act *as* claim or conjecture, and to acknowledge frankly the *identity of the claimant or conjecturer.*

III

Since being a claimant seems to me to be central to the genesis and phenomenology of the things we call mental illnesses, and since being a claimant is obviously quite peripheral to the genesis and phenomenology of the things we call bodily illnesses, I find it astonishing that this plain fact and basic distinction has been so neglected, not only by psychiatrists but also by philosophers. To illustrate this point, let us consider the situation in which a bodily disease may be discovered and diagnosed without any prior claims of illness by or about the 'patient.' On a routine medical examination for admission to college or entrance into the armed forces it is discovered, on the basis of tests of the subject's urine or blood, that he has diabetes or latent syphilis or leukemia. The subject does not claim that he is ill. No one (prior to the examination) has claimed that he is ill. In short, his illness is diagnosed completely independently of any such claims.

Is it possible to discover mental illness in a person in a similar situation and in a similar way? Clearly, it is not. The diagnosis of mental illness depends wholly on what the subject says about himself, or what others say about him. Moreover, the things that are reported to medical authorities in such contexts are in the nature of claims, some of which may be verifiable by others, but some of which may, by their very nature, not be. I do not see how we can confront the problem of the meaning of 'mental illness' without coming to grips with these typical 'psychiatric claims' – that is, with the sorts of assertions that have historically led, and that often continue to lead, to the diagnosis of mental illness.

In the case of an examination for the draft, the subject may tell the doctor that he wets the bed or that he is a homosexual. These are claims. The assertion about bedwetting is a claim that the subject may be able to verify. The claim about homosexuality may also be verifiable (for example, by a previous arrest for it), or it may be just as unverifiable as the claim of heterosexuality (which the subject would not be expected to verify). In the absence of such claims, mental illness cannot be diagnosed. This, it seems to me, is its most essential characteristic.

Let us pursue this matter of claims a bit further by inspecting some of the historical claims that have given rise to diagnoses of mental illness. These claims have, in the main, been of two types: namely, having pains in the absence of lesions legitimizing them, and not having pleasures in the presence of laws legitimizing them. In traditional psychiatry, claimants of the first type have been classified as suffering from such 'mental diseases' as hysteria, hypochondriasis, and neurasthenia, whereas claimants of the second type have been classified as exhibiting such 'mental symptoms' as frigidity and impotence.

Why do we categorize persons exhibiting such behaviors as sick? Because their alleged diseases are similar to diabetes, or because we want to justify calling the claimants 'patients?' As I have remarked elsewhere, the difference between holy water and ordinary water lies not in the water, but in the priests; similarly, the difference between the claims of so-called mental patients and the claims of other people lies not in the claims, but in the medical profession [12, 14]. Thus a woman who is sexually unresponsive to her husband (or other men) is frigid, and frigidity is a paradigmatic symptom of a mental illness. But an orthodox Jew who is alimentarily unresponsive to a ham sandwich (or other pork products) is not said to exhibit any 'symptoms' of any 'mental illness.'

I shall now try to amplify these observations by considering an actual historical paradigm of a mental illness.

IV

Except for what used to be called 'madness' and is now called 'psychosis,' the single most important mental illness, from an historical point of view, is hysteria. This is why I chose it as my model in *The Myth of Mental Illness* [7]. Here I want to consider briefly – but in sufficient detail to illustrate my argument concerning the role of claims and justifications in the very definition of mental illness – a book on hysteria written in 1917 by the great French neuropsychiatrist Joseph Babinski (1957–1932).

Hysteria, asserts Babinski, quoting with approval Ernest Laségue, 'has never been defined and never will be' ([1], p. 17). Why? Not, as it might seem, because hysteria is not a disease but is nevertheless defined as one, but because the term has been used too loosely. The remedy Babinski recommends is, therefore, to distinguish among the various things that have been placed in this category in the past, to retain some, and to reject others:

> . . . it was impossible in former times to define the conditions comprised under this title. Nowadays, it is different. The isolation of the group of phenomena which may be called indifferently hysterical or pithiatic is an accomplished fact. They possess characteristics which belong only to themselves, which are absent in all other morbid states, and which therefore constitute the elements of the definition of hysteria, which I have set forth as follows ([1], p. 17):

I hope the reader is now waiting breathlessly to hear, at last, a 'definition' of hysteria. Babinski is going to tell us what hysteria *is*, just as if it were an object or an event. He is going to 'explain' it, not 'justify' it. Or so he claims:

> Hysteria is a pathological state manifested by disorders which it is possible to reproduce exactly by suggestion in certain subjects and which can be made to disappear by the influence of persuasion (counter-suggestion) alone ([1], p. 17).

Here, then, is one of the leaders of turn-of-the-century French neuropsychiatry 'explaining' hysteria by calling it 'a pathological state manifested by disorders . . . ' – a 'definition' that is, in fact, a justification for his particular way of labeling certain kinds of actors.

Actually, Babinski defines hysteria as acting: he tells us – as I would paraphrase it – that there are two kinds of actors, legitimate and illegitimate. Legitimate actors perform on the stage. Producers and directors ask them to play this role or that, and to cease playing them. We do not call the director's communications 'suggestion' and 'counter-suggestion;' nor do we call the actors' performances 'pathological states.'

Illegitimate actors perform off stage. Psychiatrists ask them to play this role or that, and to cease playing them. We call the psychiatrists' communications 'suggestion' and 'counter-suggestion,' and the players' performances 'pathological states.'

I submit it is as simple as that. Stage actors are legitimate performers; off-stage actors are hysterics. Or, we might put it differently, as follows. Some people act sick and doctors accept their act as legitimate: they are the bodily or medically sick patients. Other people act sick and doctors do not accept their act as legitimate: they are the mentally or psychiatrically sick patients. *Mutatis mutandis*, children born to married women are legitimate; those born to unmarried women are illegitimate. In the latter case, we recognize, however, that the 'illegitimacy' of the child is actually a moral judgment about the mother; whereas in the former case we believe that the 'illegitimacy' of the act is a 'symptom' of a medical disease in the actor.

Indeed, Babinski is so naive about his medicalizing of conduct that he candidly admits that he is identifying a disease characterized by the fact that a person can be talked out of having it. He writes:

> There might even be some advantage in abandoning the use of the term hysteria, which in its etymological sense is in no way suitable for any of the phenomena under consideration. . . . I have proposed the substitution of the term 'pithiatism', from [the Greek for] 'I persuade' and 'curable,' which expresses one of the fundamental characteristics of these symptoms, viz., the possibility of being cured by the influence of persuasion ([1], p. 17).

By 'cured,' Babinski here refers simply to one person telling another to stop doing something, and the latter complying. I could not imagine anything more unlike what is entailed in curing a real disease, such as syphilis.

Having justified hysteria – renamed 'pithiatism' – as an illness, and the 'pithiatic' patient as a sick person 'cured' by 'counter-suggestion,' Babinski has fulfilled his duty as psychiatric patriot: he has conquered an area of personal conduct for pathology. He is ready, next, to fulfill his duty as regular patriot: he must now protect France from pithiatism. With boundless pride he quotes a recommendation for which he was evidently partly responsible:

> . . . on October 21, 1915, the [French] Neurological Society sent a recommendation to the Under Secretary of State for the Sanitary Service to the effect that '*no soldier at the present time, under any circumstances, with a psycho-neurosis should be brought before a medical board with a view to discharge from the army*.' (Italics in the original) ([1], p. 229).

On page 17, hysteria is a 'pathological state,' justifying its annexation to medicine; on page 229, it is a condition that under no circumstances justifies discharging the 'patient' from the army. In his recommendation concerning compensation for illness, Babinski reiterates a view that supports a policy quite inconsistent with viewing hysteria as an illness, but quite consistent with viewing it as an act (which may be called malingering, simulation, claiming to be ill, or playing patient):

At a recent meeting of the representatives of the neurological centers (December 15, 1916) to discuss the subject of 'Discharges from the Army, Disabilities and Allowances in the Neuroses,' the questions with which we have dealt in this book were discussed from the point of view of a medical board, and the conclusions of a report drawn up by one of us were adopted to serve as the groundwork for the next edition of *The Ready Reckoner of Disabilities.* They are of interest to all medical officers, and may therefore be reproduced here. They are as follows: 1. For purely hysterical or pithiatic disorders: no discharge nor allowances ([1], p. 234).

Babinski's dilemma, his apparently complete obliviousness to it, and the consequent glaring inconsistencies in his claims about hysteria continue to be very relevant to the problems facing contemporary psychiatrists: like Babinski, they too would like to expand or contract the category of disease according to whether they seek their rewards from medical or military (or other) authorities.

As a medical patriot, Babinski saw his duty as declaring as many men as possible to be sick; whereas as a French patriot, he saw it as declaring as many men as possible to be fit for military service. He thus exemplifies the 'great' psychiatrists whom he imitated, and who in turn imitated him: he could conceive of no intellectual or moral duty other than serving Medicine or the State.

V

Babinski was, of course, neither the first nor the last to resort to the rhetoric of madness to conceal denigration as diagnosis. While I have been unable to trace this practice to its earliest origins, I have, in my search for these origins, found some remarkable examples of it. Here is one from the fourth century:

Emperors Gratian, Valentinian, and Theodosius Augustuses: An Edict to the People of the City of Constantinople.

It is Our will that all the people who are ruled by the administration of Our Clemency shall practice that religion which the divine Peter the Apostle transmitted to the Romans ... We command that those persons who follow this rule shall embrace the name of Catholic Christians. The rest, however, whom we adjudge demented and insane, shall sustain the infamy of heretical dogmas, their meeting places shall not receive the name of churches, and they shall be smitten first by divine vengeance and secondly by the retribution of Our own initiative, which We shall assume in accordance with the divine judgment ([4], p. 440).

This prescription – ostensibly describing as 'demented and insane' those who reject the Catholic faith – was issued 'on the third day before the kalends of March, at Thessalonica, in the year of the fifth consulship of Gratian Augustus and the first consulship of Theodosius Augustus' (February 28, 380) ([4], p. 440). It forms a part of the *Codex Theodosianus*, or the *Theodosian Law Codes*, published in their final form in 438, and constituting then the essential body of Roman Law.

In keeping with this law, in the fourth century 'heretics' were categorized as 'demented and insane' by the priests, and persecuted 'in accordance with divine judgment.' *Mutatis mutandis*, in the twentieth century, 'heretics' are categorized as 'psychotic and schizophrenic' by psychiatrists, and persecuted 'in accordance with diagnostic judgment.'

Moving from the sublime to the ridiculous, or at least from the theological to the theatrical, here is an illustration from the recent literature of my contention that psychiatric diagnostic terms are sadly lacking in descriptive content. Peter Shaffer's play *Equus* was a great theatrical success in America, despite an attack on it by a Harvard psychiatrist. Sanford Gifford, the psychiatrist, characterized the play in the *New York Times* as a 'fictitious piece of psychopathology. The basic ingredient in the central character's syndrome is hysteria . . . ' ([5], p. 11). He had some quite nasty things to say about the play which are fortunately not relevant to our present concerns.

I mention *Equus* and Gifford's attack on it because in September, 1975, the magazine *Frontiers of Psychiatry*, published by the Roche pharmaceutical company, devoted a whole issue to the psychiatric implications of the play. In it, there appeared the following remarkable footnote (in small print):

> Since the American Psychiatric Association task force that is preparing the third edition of the *Diagnostic and Statistical Manual* has announced its intention to delete hysteria as a diagnostic term from Category VII, Hysterical Disorders, Roche Report informally questioned four psychiatrists (one, a specialist in diagnostic terms; another, a child analyst), three psychologists, and four social workers, all of whom had seen *Equus*, and asked them how they would diagnose the boy as he presented at the hospital. There were nine different opinions, ranging from 'adolescent episode' through 'transient psychosis' to 'schizophrenic break.' Two psychiatrists took the Fifth Amendment ([1], p. 6).

What should we make of this? We might use this incriminating poll to justify our contemp to psychiatry – which would be a good thing, so far as it went; but it would trap us into adopting the same sort of behavior toward psychiatrists that psychiatrists exhibit toward patients. Can we do better? I think so – by concluding that psychiatrists are, in fact, crypto-priests, and that their job is to bless and to damn. Holy water is holy not because of the kind of water it is, but because a priest has blessed it. A schizophrenic patient is schizophrenic not because of the sort of person he is but because a psychiatrist has damned him [12, 14].

Suppose that, in accordance with the 'diagnostic' manual of the *Theodosian Law Codes*, a person had been declared 'insane.' Clearly, that term did not, and was not intended to, describe his beliefs: such a person might have been a believer in many Roman gods, in one Jewish god, or in no god. But the theological diagnosticians of that age were not interested in classifying his *beliefs*; they were interested in classifying *him*. They were not interested in *understanding* his beliefs; they were interested in *justifying* their own condemnation and destruction of him.

Similarly, when a playwright now offers us a dramatic encounter between a modern madman and his mad-doctor, and when psychiatrists are asked to 'diagnose' the 'patient,' it is hardly surprising that they make nearly as many diagnoses as there are diagnosticians. After all, the diagnostic terms they are supposed to produce, in response to such a query, are all the synonyms of the ancient term 'insane:' they are all psychiatric terms of abuse, not descriptive terms referring to any actual human behavior.

Ironically, the real description of behavior in plays such as *Equus* – or *Hamlet*, or *King Lear* – is in the plays themselves. The work of art – the play, the performance – is, as it were, the 'science.' The psychiatric 'explanation' of such behavior thus not only lacks any genuine descriptive content, but is actually a distortion, and a veritable destruction, of precisely the sorts of 'data' that constitute the raw materials of science. Indeed, the 'denaturing' or 'falsification' of the 'data' of direct observation and plain reporting is what psychiatry is all about. A few remarks about this process may be in order here.

VI

What is industrial alcohol? It is 'denatured' alcohol – that is, pure alcohol made unfit for human consumption.

I submit that 'mental illness' stands in the same sort of relationship to human behavior as industrial alcohol stands to ethanol. Distilleries produce alcohol that is, more or less, chemically pure. It requires the active intervention of human beings to make this 'natural' product 'unnatural:' hence the apt name for it, 'denatured alcohol.' Similarly, human beings 'produce' behavior. They act. They speak. It requires the active intervention of human beings to make these 'natural' products 'unnatural:' hence the revealing names for such acts – 'perversion,' 'delusion,' 'psychosis.'

Let us take a simple example. A man declares that he is Jesus. What shall we make of it?

First, we might take it for just what it is: a person asserting that he is the Savior.

Second, we might respond to it plainly, matching it against our own knowledge of the world: a person asserting a false identity; in short, a liar.

Third, we might respond to it psychiatrically, matching it against our knowledge of psychiatry: a person displaying a delusion; in short, a psychotic.

The question we must face is this: is the psychiatric account of such behavior a more abstract and 'scientific' description, or even explanation, of such behavior, or is it the distortion and redefinition of it? I have argued that it is, overwhelmingly, the latter. I thus hold that learning psychiatry is largely a matter of learning to see human behavior – perhaps even the whole world – through the distorting lenses of this fake science. People who learn this lesson – whether as professional psychiatrists or as laymen 'educated' in mental health – thus learn, first of all, that there are two kinds of behavior, and two kinds of people, in the world – mentally healthy and mentally sick. Having learned that lesson, they are ready to tackle the problem of 'understanding mental illness.'

This approach and perspective are exemplified by virtually every psychiatric publication. A recent article entitled 'Some Myths about "Mental Illness",' by Michael S. Moore, a professor of law at the University of Kansas is typical. One of Moore's conclusions, offered at the end of a long argument based on a mixture of a few assertions I have made and of many I have not, is this:

> No one merits society's condemnation or punishment unless they are morally blameworthy, and no one is blameworthy if he acts as he does because of his mental disease. Szasz's conclusion is reminiscent of the Erewhonian practice of punishing the ill, and evokes in most of us the same distaste ([3], p. 1495).

I should like to note that Moore actually packs two mistakes into a single sentence:

(1) By asserting that no one merits punishment who is not blameworthy, he implies that by treating 'mentally ill' people as 'mentally ill' we do not punish them; but by treating them as if they were not 'mentally ill' we do, *ipso facto*, punish them.

(2) He also asserts that some people act as they do *because* of their 'mental disease.'

I disagree with the second assertion. But that is not the point here. The point is that Moore adduces no evidence in support of the view that some people act as they do because of their mental illness, and others because of another 'cause.' What, one wonders, could that 'cause' be? Free will?

As to the second assertion, Moore surely must have heard, before Andrei Sakharov announced it in his Nobel lecture, that 'Worst of all is the hell that exists in the special psychiatric clinics' [6]. Yet Moore claims that by opposing involuntary psychiatric interventions I propose to 'punish the ill.' This is a typical instance of the breakdown of language between the adherents to psychiatric true belief and psychiatric agnostics. The true believers assert, with Moore, that:

> Since mental illness negates our assumptions of rationality, we do not hold the mentally ill responsible. It is not so much that we excuse them from a *prima facie* case of responsibility; rather, by being unable to regard them as fully rational beings, we cannot affirm the essential condition to viewing them as moral agents to begin with. In this the mentally ill join (to a decreasing degree) infants, wild beasts, plants, and stones – none of which are [*sic*] responsible because of the absence of any assumption of rationality ([3], p. 1496).

To which I reply: If this is not using language to dehumanize and destroy persons then I do not know what is. Moore here reasserts the proposition that some individuals who seem to be persons are, in fact, not; they ought to be classed with 'wild beasts, plants, and stones.'

There is, of course, nothing new about this idea. Nor is there anything new about the irreconcilable conflicts such 'religious' controversies generate. On this occasion, it must suffice to state, or restate, the terms of the controversy. The supporters of the concept of mental illness claim that 'madmen' are like plants and we should, in order to be 'good' to them, treat them as if they were plants. I say that 'madmen' are persons and that we should treat them as if they were.

VII

The view that 'mental illnesses' have to do with claiming and justifying, rather than with diseases and treatments, helps to explain why psychiatry has traditionally been linked to the law in ways quite unlike the rest of medicine. Like psychiatry, the law deals with conduct and with the justification of conduct.

Indeed, Anglo–American law is premised on, and displays, precisely the sort of understanding of human acts which I have sketched – and which is obscured by psychiatry; and it is aimed at resolving conflicting claims between persons fairly and consistently – a process which psychiatry renders unfair and capricious. In both civil and criminal trials, the arbiters assume that plaintiff and defendant, prosecuting attorney and defense attorney, each presents different claims and conjectures about why the protagonists in the judicial drama acted as they did. It is up to the jury to develop its own conjectures, whose consequences the court then imposes on the litigants. The jury, or court, does so not because it is more intelligent or more honest than the participants in the litigation, but because it is more neutral than they are, and because it has the authority and the power to do so.

Psychiatric 'expert' testimony distorts, and indeed destroys, this judicial arrangement in which facts are reasonably well demarcated from opinions. The reason for this is that in so far as the psychiatrist testifies about why a person acted as he did, he offers a conjecture which, however, is widely defined and accepted as a cause. This is epitomized by the belief – now authoritatively accepted as scientifically 'correct' or 'true' – that some people kill because they hate their victims, others because they want their money, and still others because they 'have schizophrenia.' Mental illness as a cause, and murder as a product of it, must thus be seen for what it is: not just a mistaken idea, but the manifestation of the judicial acceptance – and indeed acclaim – of the psychiatrist as scientist of the mind.

I maintain that it is precisely this acceptance of psychiatry as a legitimate science that is responsible for two closely interrelated phenomena: the corruption of the administration of justice in the courtroom; and the confusion of the nature of human behavior in the classroom. Each of these processes has, of course, consequences that extend far beyond the actual locations of these 'rooms.'

Suppose that astrology were accepted as a legitimate science, and that astrologers were allowed to testify in court the way psychiatrists now are. We might then have the spectacle of one set of astrologers testifying that, because of the constellation of planets on the night of a particular murder, the accused was not responsible for his act; while another set of astrologers, basing its 'expert opinion' on ostensibly the same 'data,' would testify that the accused was responsible for his act. If, despite this kind of astrological testimony, astrology would continue to enjoy the unqualified support of the scientific community, then astrological 'theories' would seriously interfere with our understanding of the nature of human behavior. Indeed, this is no idle analogizing. We saw this very thing happen with theological theories concerning witchcraft and their impact on both the popular and 'scientific' understanding of human behavior [9].

The fact, then, that psychiatry is an accepted field of science – it is taught in medical schools next to subjects such as biochemistry and physiology; it is supported by the government; and it is accepted in courts of law as if it had 'methods' similar to those of ballistics or toxicology – exercises a significant influence on the very 'observations' about which its practitioners supposedly possess 'expert' knowledge. The concept of 'mental illness' itself, as I tried to show elsewhere [7], embodies and epitomizes this fatal prejudgment – namely, that 'scientists' have identified a phenomenon which they, and we, must try to understand. But the term 'mental illness' does not identify any clearcut phenomenon or class of phenomena; instead, it is a piece of self-justificatory rhetoric that precludes our understanding of the very problems we are supposed to be trying to understand.

VIII

In actuality what determines whether a person's sanity is called into question is not what he, as an actor, does, but rather what the authorities, as his audience, think about it. As a rule, behavior deemed to be good or desirable is accepted without further justification, whereas behavior deemed to be bad or undesirable is not. Hence it is that people now turn to psychiatrists to explain why someone committed a bad deed, but do not turn to them to explain why someone committed a good deed. Thus, confronted with a person who destroys life, say by killing many people, the most natural thing to do now is to ask psychiatrists to 'explain' why he did so; and the most 'scientific' thing is to accept that he did so because of his 'mental illness.' However, confronted by a person who preserves life, say by discovering how to prevent or cure a disease, the most absurd thing to do would be to ask a psychiatrist to explain why he did so; and the most 'unscientific' thing would be to accept that he did so because of his 'mental illness.'

In other words, although we speak about medical diseases when we are confronted by deviations from biological norms, and about mental diseases when we are confronted by deviations from behavioral norms, only in the former case do we designate deviations from the norm as 'diseases' regardless of the direction the deviation takes. Thus, whether a person has too many white corpuscles or too few, he is said to have a disease. This is not true for behavioral 'abnormalities,' virtually all of which designate deviations from the norm in one particular direction – that is, toward the immoral or illegal. For example, psychiatrists consider persons who are more wicked than the norm 'sick' and call his disease 'psychopathy,' but they do not consider persons who are more virtuous than the norm sick and have no 'disease' corresponding to psychopathy to 'explain' their behavior. This discrepancy between our attitude toward 'explaining' good and bad deeds should alone suffice to show how pervasively we have confused justifications of human behavior with their explanations.

Although all this might be obvious to some, the fact remains that the whole history of institutional psychiatry is characterized by, and is the consequence of, the experts' insistence that what needs attending to is not the public or professional acceptance or rejection of acts, leading to their justification or nonjustification – but the actor's sanity or insanity [11]. By consistently asking the wrong questions – 'wrong' in the sense that they are psychiatrically self-serving, rather than phenomenologically significant – psychiatrists, and all who have accepted their premises, could come up with nothing but wrong answers.

IX

The fact that the phrases found in textbooks of psychiatry are in the main justifications rather than explanations goes a long way toward accounting for the stubborn disjunctions between avowed claims and actual conduct so characteristic of the behavior of both mental patients and psychiatrists. For example, a person, likely to be diagnosed schizophrenic, may declare: 'I am the Messiah, God commands me to save the world, and to do so I must kill so-and-so.' The 'patient's' putative aim is to 'save the world,' to do good. However, his actual conduct, as judged by the recipients of his benevolence, is deemed to be 'dangerous' and 'harmful,' with consequences all too familiar.

The situation with respect to the psychiatrist is much the same, with the roles reversed. He declares: 'I am a doctor, my medical training and ethic command me that I help sick people, and to do so I must treat so-and-so for his schizophrenia.' The doctor's putative aim is to help the 'patient,' to treat him for his 'disease.' However, his actual conduct, as judged by the recipients of his benevolence, is deemed to be not treatment but torture, with consequences again all too familiar.

Such disjunctions between putative aims and actual performances cannot long stand unresolved. In the modern world they are resolved, at least in the areas I am here considering, by the simple expedient of substituting authority for evidence, power for compassion, force for reason. Thus, when the majority, the government, and science – through its duly appointed agents and agencies – declare, as they do in the case of mad-doctoring, that its actual performances are the same as, or closely approximate, its putative objectives, the disjunction between the 'doctor's' self-serving aims and his other-damaging acts is instantly resolved. Indeed, it is better than resolved: it is defined out of existence.

Similarly, when the majority, the government, and science – through their duly appointed agents and agencies – declare, as they do in the case of madness, that both its putative objectives and actual performances are the meaningless 'symptoms' of a medical disorder, the disturbing disjunction between the self-serving aims of the 'madman' and his other-damaging acts is again instantly resolved. And again it is better than resolved: it is defined out of existence. Henceforth, both of these disjunctions can be recognized and addressed only at the risk of insulting established professional beliefs and practices, and incurring the risks customarily accompanying such behavior.

In my opinion, herein lie the fundamental ideological, economic, and political sources of the difficulties that face the contemporary student of psychiatry. Countless psychiatric principles today are based on, or articulate, deliberate deceptions – such as calling people who reject medical help 'patients' and the buildings in which they are imprisoned 'hospitals.' And countless psychiatric practices today consist of nothing but crass coercions – such as the incarceration of innocent persons under psychiatric auspices called 'mental hospitalization.' These dramatic disjunctions between putative objectives and actual performances, pervasive of all institutional psychiatry, are now supported by both church and state, law and science. Accordingly, the psychiatric scholar's first task must be to re-assert the evidence of his naked eyes and ears. For it is of little use to explain, justify, or modify policies that linguistically entail events as facts which are actually frauds, and that morally authenticate aims as medical and technical albeit actually they are moral and political.

X

In conclusion I should like to restate one of the central arguments that I have been making about psychiatry for the past twenty years or more, and that I have also made in this essay.

I do not assert, as some of my critics claim, that psychiatry is not a science because it deals with non-existent things, such as 'mental illnesses.' I assert that psychiatry is not a science because its practitioners are basically hostile to the ethic of truth-telling. Why should this be so? Assuredly not because psychiatrists are bad people; they are, on the whole, no better or worse than other people. Instead, it is because the ethic of truth-telling, as it is institutionalized in modern science, is not found outside of the narrow border of science. Politicians and priests, lawyers and theologians are all important and respectable members of society; but the crafts they practice are not based on truth-telling.

If, as I claim, truth-telling is the essential ethical-linguistic 'method' of science, what is the corresponding 'method' that characterizes such professions or groups as priests, politicians, and psychiatrists? It is the telling of literalized metaphors, strategic myths, and even of calculated mendacities – which are designed to advance not only the interests of the speaker but also those of his 'club.' Whatever else priests, politicians, and psychiatrists may say or do, they must, to remain in good standing, protect and promote the greater glory of God, the fatherland, and 'mental health.' This is why, strictly speaking, there can be no sciences of priesthood, politics, or psychiatry; there can be only criticisms of them.

State University of New York,
Upstate Medical Center,
Syracuse, New York

BIBLIOGRAPHY

1. Babinski, J. and Froment, J.: 1918, *Hysteria or Pithiatism*, trans. by J. D. Rolleston, University of London Press, London.
2. Carlyle, T.: 1890, *Sartor Resartus*, Home Book Co., New York.
3. Moore, M. S.: 1975, 'Some Myths about "Mental Illness"', *Archives of General Psychiatry* **32**, 1483–1497.
4. Pharr, C.: 1952, *The Theodosian Code and Novels and the Sirmondian Constitutions*, Princeton University Press, Princeton.
5. Point of View: 1975, 'Playwright Peter Shaffer Raps American Normalcy, Conformity, Psychotherapy', *Roche Report: Frontiers of Psychiatry* **5**, 1–2, 5–11.
6. Sakharov, A.: 1975, 'Excerpts from the Nobel Lecture', *The New York Times*, Dec. 13, p. 6.
7. Szasz, T. S.: 1961, *The Myth of Mental Illness*, Hoeber-Harper, New York.
8. Szasz, T. S.: 1970, *Ideology and Insanity*, Doubleday Anchor, Garden City, New York.
9. Szasz, T. S.: 1970, *The Manufacture of Madness*, Harper and Row, New York.
10. Szasz, T. S.: 1970, 'Mental Illness as a Metaphor', *Nature* **242**, 305–307.
11. Szasz, T. S.: 1973, *The Age of Madness*, Doubleday Anchor, Garden City, New York.
12. Szasz, T. S.: 1973, *The Second Sin*, Doubleday Anchor, Garden City, New York.
13. Szasz, T. S.: 1974, *Ceremonial Chemistry*, Doubleday Anchor, Garden City, New York.
14. Szasz, T. S.: 1976, *Heresies*, Doubleday Anchor, Garden City, New York.

BARUCH BRODY

SZASZ ON MENTAL ILLNESS

Dr. Szasz tells us that he will argue that the concept of mental illness causes continuing difficulties because it is a prescriptive concept being used as though it were a descriptive concept, a justificatory concept disguised as an explanatory concept. If his arguments were correct, then he would certainly have spotted a fundamental difficulty with the current use of that concept. I do not know whether or not Dr. Szasz's claim is ultimately correct; all that I shall try to show in my remarks is that Dr. Szasz's arguments for this claim fail.

Two preliminary points. First, by his own account of what he is trying to establish, the most that Dr. Szasz can hope to show is that there is something wrong with the current use of our concepts of mental illness, not that there is something inherently wrong with those concepts. That stronger thesis, which I believe Dr. Szasz holds, would require a different argument. Secondly, there is a confusion in Szasz's account of how the concepts of mental illness are misused. He sometimes says that they are mistakenly used as descriptive concepts, but on other occasions he says that they are mistakenly used as explanatory concepts. Now, in general, a description is not equivalent to an explanation. I may describe what has happened by saying that John killed Frank. That description provides us with no explanation of what has happened. It would seem, therefore, that descriptive concepts (concepts used in descriptions) are not necessarily identical with explanatory concepts (concepts used in explanations).

In light of these two points, it would probably be best to take Dr. Szasz's thesis to be: the current use of concepts of mental illness creates difficulties because it uses a prescriptive concept (one used to justify certain behavior towards people) as though it described or explained the behavior and state of the people in question.

I. MENTAL ILLNESS AND CLAIMANTS

In Sections II and III of his paper, Dr. Szasz sets out an argument that diagnoses of mental illness are necessarily claim-based diagnoses and are, therefore, an attempt to justify certain types of treatment rather than a description and explanation. I frankly do not see either why he accepts his premises or why he thinks that the conclusion follows from them.

H. T. Engelhardt, Jr. and S. F. Spicker (eds.), Mental Health: Philosophical Perspectives, 251–257.

II. MENTAL ILLNESS AND NON-NORMAL BEHAVIOR

At a number of points in his paper, Dr. Szasz argues for his thesis by appealing to the relation between our classifying a type of behavior as a form of mental illness and our evaluating that behavior. The following passage is typical:

In other words, although we speak about medical diseases when we are confronted by deviations from biological norms, and about mental diseases when we are confronted by deviations from behavioral norms, only in the former case do we designate deviations from the norm as 'diseases' regardless of the direction the deviation takes. Thus, whether a person has too many white corpuscles or too few, he is said to have a disease. This is not true for behavioral 'abnormalities,' virtually all of which designate deviations from the norm in one particular direction – that is, toward the immoral or the illegal. For example, psychiatrists consider persons who are more wicked than the norm 'sick' and call his disease 'psychopathy,' but they do not consider persons who are more virtuous than the norm sick and have no 'disease' corresponding to psychopathy to 'explain' their behavior. This discrepancy between our attitude toward 'explaining' good and bad deeds should alone suffice to show how pervasively we have confused justifications of human behavior with their explanations ([1], p. 247).

I find in this argument, as in the last, difficulties both with the premises and the transition to the conclusion.

To begin with, it seems just false that only deviations from the norm in the direction of immoral or illegal behavior are considered diseased behavior patterns. The person who washes his hands two hundred times a day behaves in a pattern that we describe and/or explain by reference to concepts of mental illness although his behavior is neither immoral nor illegal. The person who is terrified of enclosed spaces and refuses to enter elevators behaves in a pattern that we describe and/or explain by reference to concepts of mental illness although his behavior is neither immoral nor illegal.

Secondly, it seems just false that any deviation from a biological norm is a bodily disease. People who are considerably taller than normal, people with unusually shaped noses, people with unusually brilliant red hair, are all people who deviate from biological norms but are not diseased. There are, indeed, many types of deviations from biological norms that are physical assets rather than diseases. Examples that immediately come to mind are unusual strength, especially keen sight, etc.

These remarks raise, of course, a very fundamental problem about the concept of a disease. What type of deviation from the norm constitutes a disease? (In fact, can there be a disease that is a norm?) Is it a function of the survival disvalue of the deviation? Is it a function of some essentialist conception of a well-functioning human being? Is it a function of some social disvalue of the deviation? These questions arise for both mental and physical diseases, and I cannot now attempt to deal with them. All that I can say at this point is that Szasz's account, which is the premise of his argument, is clearly wrong.

Suppose, however, that there is some element of truth in Szasz's account, suppose that we only describe something as an illness if we disvalue it. Does that entail that concepts of mental illness are only justificatory of our responses to behavior and are not explanations and/or descriptions of it as well? I do not see why it is supposed to entail the conclusion. Szasz presumably has in mind the following: given that we have picked out as diseased only socially objectionable forms of behavior, we clearly are doing so to justify certain treatments of the people in question. Suppose that this is true; it still does not follow that the concepts in question cannot also be descriptive and/or explanative.

This leads me to my second major conclusion about Dr. Szasz's arguments. Much as he is wrong about the relation between mental illness and behavioral patterns, so he is confused about the relation between the prescriptive and the descriptive and/or explanative. These concepts are not mutually exclusive; there is no reason why a concept cannot have both descriptive and prescriptive content. Consider the familiar example of the concept of an apple. A good apple must satisfy certain descriptive criteria and yet calling something a good apple prescribes choosing it if one wants a good apple. There is no reason, similarly, why mental-illness concepts cannot be both descriptive and prescriptive.

At one point, Szasz sets forth the following argument for doubting that concepts of mental illness have any descriptive content:

> Similarly, when a playwright now offers us a dramatic encounter between a modern madman and his mad-doctor, and when psychiatrists are asked to 'diagnose' the 'patient,' it is hardly surprising that they make nearly as many diagnoses as there are diagnosticians. After all, the diagnostic terms they are supposed to produce, in response to such a query, are all the synonyms of the ancient term 'insane:' they are all psychiatric terms of abuse, not descriptions referring to any actual human behavior ([1], p. 243).

There is, I think, something of a point here. Psychiatric terms are not as well-defined as most medical terms; there certainly are few operational definitions for such terms. But, of course, none of this entails that they do not refer to any actual human behavior. It shows at best that their mode of reference is indirect, and one of the fundamental lessons of modern philosophy of science is that theoretical terms refer to observational data only in this indirect fashion. So nothing much follows from what is right in this last remark of Szasz's.

III. ON BABINSKI

In advancing his argument, Dr. Szasz devotes much space to Babinski's account of hysteria as illustrative of his thesis. I do not see what it is supposed to illustrate, but let me in any case point out a number of further confusions in Szasz's treatment of that case.

Babinski's definition of hysteria is clearly inadequate, for it fails to deal with Szasz's counter-example of the actor. Still, what follows? Our not treating the actor the way we treat the patient can be justified by the context surrounding that behavior, and that context also provides the descriptive difference between the actor and the patient. So unless we are not allowed to count contextural features, and all concepts of mental illness certainly must involve these features, we can provide adequate descriptive criteria for hysteria.

Szasz also finds it very revealing that Babinski defines hysteria as a disease that one can be talked out of. He writes:

> By 'cured,' Babinski here refers simply to one person telling another to stop doing something, and the latter complying. I could not imagine anything more unlike what is entailed in curing a real disease, such as syphilis ([1], p. 240).

There is no doubt that curing a case of hysteria is unlike curing a case of syphilis (as, no doubt, contracting a case of one is very unlike contracting a case of the other). But why is that supposed to be problematic? The crucial resemblance between the two cures is the modification of the condition of the patient so that the disease is no longer present. As long as the concept of the disease is satisfactory (and Szasz has yet to show anything wrong with it), the concept of a cure will be acceptable.

Finally, Dr. Szasz is extremely unhappy about Babinski's going on to argue against medical discharges based upon hysteria. How, he seems to ask, can Babinski really treat hysteria as a disease and yet argue against using it as a basis for discharging soldiers? A number of arguments actually come to mind: (1) the difficulty of telling real cases of hysteria from sham cases; (2) France's desperation required it to use diseased soldiers; (3) the disease may still allow soldiers to fight adequately. I do not say that Babinski was right. In fact, I do not know what led him to his conclusion. All that I want to claim is that Szasz has failed to establish an inconsistency in Babinski's approach to hysteria.

IV. CONCLUSION

Dr. Szasz may well be right. But if he wants to show that he is right, he will have to present more substantial and careful arguments than the ones that he does present.

Rice University,
Houston, Texas

REFERENCE

1. Szasz, Thomas S.: 1977, 'The Concept of Mental Illness: Explanation or Justification', in this volume, pp. 235–250.

SECTION VI

Round-Table Discussion

REAPPRAISING THE CONCEPTS OF MENTAL HEALTH AND DISEASE

ROUND-TABLE DISCUSSION

H. TRISTRAM ENGELHARDT, JR.

CHAIRMAN'S REMARKS

To speak of mental health is to talk either as though minds could have difficulties in partial independence of their bodies, or as though a clear distinction could be drawn between the life of the mind and the life of the body. And surely minds can be distinguished from bodies, and bodies from minds, in the sense that we can speak of one without implying the other. This, though, does not mean that the terms 'mind' and 'body' identify separate substances.

Distinctions such as between mind and body, between mental physical health and disease, are always made at a price. After all, it is persons who are sick or healthy, not simply minds or bodies. It is persons who have 'presenting complaints,' as Toulmin indicates, which the health care professions then characterize as somatic or mental illnesses. Such categorizations require judgments as to the relative usefulness of styling a particular condition as physical or mental. Consider, for example, the category 'psychosis with intercranial neoplasma' listed in the diagnostical statistical manual of the American Psychiatric Association [1]. The cause is somatic, but it is classified as a psychiatric syndrome because of the usefulness of enlisting the care of mental health workers for the amelioration of the symptoms. Even diseases such as schizophrenia which at present have no clearly established physical etiology may in the end be viewed so that the choice of treatment will be based primarily on the usefulness of having the accent fall either upon mental or physical elements of human life (e.g., schizophrenia is for the most part treated somatically, even if it may be psychologically caused). But as Toulmin and Fabrega have argued, it will always be an error to view disease simply as a physical process, for it is also a complaint of a person with a rich and complete mental life. To present only the mental or the physical dimension of a person's complaint, is usually to provide a one-sided view. So, for example, shortness of breath in congestive heart failure is a form of physiological distress – but it is experienced as psychological distress as well.

Parts of the Round Table Discussion emphasize the broad meaning of mental health which includes self-realization, self-actualization, and the acquisition of personal insight. As such, the pursuit of mental health becomes

H. T. Engelhardt, Jr. and S. F. Spicker (eds.), Mental Health: Philosophical Perspectives, 261–293.

a pursuit of intellectual virtues or, perhaps more broadly, virtues of character. There is, in this sense, merit in terming much of what is sought in the psychiatrist's or psychologist's offices as a moral quest, if one construes the moral virtues broadly, as clarity and assurance in action and perception. We are reminded, in part of the discussion that follows, of the scope of the concept of health and the special freight borne by the notion of mental health. Mental health is often taken to mean a successful adaptation to the merits and exigencies of one's environment, including one's culture – a clarity in perception and action. As a result, the search for mental health and the use of certain forms of 'therapy' take on a positive significance (i.e., as not merely the 'treatment,' the negating of a disorder). What is sought is an enrichment that would be trivialized by being termed a "cure." Thus, enterprises such as psychoanalysis may maintain a validity, even if they fail to cure, as long as the person 'treated' is enriched in vision and capacity for action.

The discussions that follow remind us that there is both a privative sense of mental health (the absence of mental diseases) and a positive sense of mental health (e.g., the presence of clarity in self-understanding and action). This ambiguity is in part problematic, for it impedes a straightforward presentation of the goals and purposes of mental health programs and therapies. But the ambiguity is also strategic. It helps to direct our attention to mental health because 'mental health' signals the full and competent life that most humans would wish to realize. It is this ambiguity which makes 'mental health' both an intriguing concept for philosophers, as well as one that can benefit from philosophical analysis.

BIBLIOGRAPHY

1. Committee on Nomenclature and Statistics of the American Psychiatric Association: 1968, *Diagnostic and Statistical Manual of Mental Disorders* (Second Edition), American Psychiatric Association, Washington, D.C., p. 6.

HORACIO FABREGA, JR.

The papers presented at the symposium during the last few days cover a wide spectrum of issues and raise a host of questions. I would like to concentrate on only a few of these. The material that I draw on will naturally be selective and will reflect my interests and biases. Failure to touch on other questions and on the work of certain of the participants merely reflects constraints of time and what I feel competent to discuss.

I would like to ask for some freedom to paraphrase some of the ideas which I feel are implicit in Dr. van den Berg's paper. I find these ideas very stimulating as well as theoretically compelling. In all fairness to van den Berg, I must say that he may very well not agree with what I say. Any controversies which my paraphrasing gives rise to, I naturally take responsibility for.

Van den Berg informs us that in his estimation, neurosis was invented or created at a certain point in recent European history. I believe that this implies that psychiatric disease – say schizophrenia or dementia praecox – was invented in Western Europe in the 19th century. A distinctive cultural context gave rise to these psychiatric entities, in other words. I also believe that this implies that a very large number of medical diseases (including 'non-psychiatric' ones) were recently *created* as well.

Now, it is obvious that to a strict biomedic – say a paleopathologist – notions such as these would be vigorously denied. For example, he would state that earlier man's remains show unmistakable evidence of bone changes analogous to those we now believe reflect cancer or leprosy! A psychiatrist would say that obviously the Greeks showed depressive diseases, schizophrenia, and epilepsy. In certain ways one cannot but agree with the claims of the biomedic. There is another side to the problem, however. It seems to me that if a historical–cultural context affects how we conceive of and orient to disease, then it may also affect its phenomenal content and certainly will affect the sick person's experiences, which together I would describe as the behavioral properties of disease. It is this behavioral whole which in my estimation is critical. I would claim that all diseases, including those we now term non-psychiatric and psychiatric, showed themselves in earlier epochs to have a constitutional whole basis or content, to use van den Berg's phrase. In my view – and also, I believe, Toulmin's view – disease, almost as a matter of definition, is a unitary 'whole' thing, involving the behavior and adaptation of the person. We as biomedics – were we allowed to visit ancient Greece, medieval England, or even contemporary non-literates – might identify different types of disease we now accept as real and judge as timeless – e.g., hypertension, schizophrenia, diabetes, etc. In this sense – with a pardon expressed on my part to Bishop Berkeley – such diseases existed then. Yet, to a metabletician (I am not sure this is a correct word) such diseases really did not exist as such! They were all, when expressed 'naturally,' compounded of physical signs and symptoms, behavioral changes, and mental symptoms, if you will. Recall that a behavioral pattern constitutes the 'natural' expression of disease, this pattern is made sense of *by us* and in doing so we reflect our presuppositions about 'disease' – what I referred to yesterday as our taxonomy.

although directed to psychiatry, are applicable to physicians in general and to the role which medicine and disease have come to play in our society.

Szasz suggests that the area which involves the so-called motives and meanings of human social behavior does not admit of a veridical rendition. That is, it is not subject to the truth-telling mode of science. What we have available, instead, are simply different claims about or versions of the 'why' of behavior. Such perspectives about the 'why' of behavior necessarily reflect the vested interests of persons and groups. Since there is no one true basis or meaning of the 'why' of human behavior, there is no justification for expecting that any one's claims about the motives of others, however expert society may judge him to be, will render a true account of behavior. Claims about mentally healthy or mentally unhealthy motives, i.e., psychiatric disease, then, simply have no warrant as scientific facts – hence should not be used as a scientific basis for justifying social policy or action.

One seems urged to acknowledge that the interpretation of behavior – the business of assigning a social meaning to the factors generating it – requires prudently opting for a set of arbitrary rules. These rules should govern the resolution of contrary claims which arise about social problems linked to behavior. Szasz seems to suggest that traditional civil-libertarian notions, including socio-legal practices and institutions, be accorded the sole power and privilege of establishing these rules – as well as criteria about evidence of rule breaking and sanctions to the rule breaker. Psychiatrists (he should say, physicians) have no logical place in this scheme – they simply cannot furnish us with a 'scientific' truth which can undermine the social meaning of behavior which all of us as citizens respond to in terms of our traditional rules and expectations. Psychiatrists' testimonies, and the idea of psychiatric disease, are not only a cop out – they are logically unwarranted.

I do not agree entirely with Szasz's thesis. I do not feel that the case for the uniqueness of non-psychiatric diseases is that strong. More specifically, I believe that psychiatric diseases are just as 'real' and 'valid' as non-psychiatric ones. This follows from my perspective about the meaning of disease. However, given the meaning of our medical taxonomy, all diseases can and will pose problems which involve responsibility for social actions. I also believe that psychiatric diseases stem from and also produce certain special paradoxes which I touched on in the paper I presented in this conference and have discussed in other papers. Szasz has pointed to a number of problems involving contemporary medical institutions and practice. In order to deal effectively with those problems, we need to keep separate six notions: (a) the illness which an individual develops – in the last analysis, a set of (negative) behaviors; (b) diseases physicians can legitimately *diagnose*; (c) *patients* who wish and seek treatment for the problems posed to them by these diseases; (d) responsibilities which doctors and patients have for each other; (e) the matter of what social institutions should do about the social actions and behaviors of persons who come to be diagnosed and/or treated by physicians, and (f) the role which physicians or any other group of experts should play in the way non-medical institutions function. Problems involving relations between and confusions about these six issues inevitably follow from the evolution of biomedicine – the metabletic revolutions of van den Berg. Conferences such as this one will help clarify these problems. I certainly leave the conference with a clearer perspective on them. I believe we owe Szasz our thanks for having sharpened these issues in the first place.

I would like to add that Toulmin's argument for what I would term a holistic, behavioral view of *all* disease (as it is handled in practice) can be used to indicate that psychiatrists are not alone in offering medical justifications of social behavior. All physicians can be held to do this when they "certify" work absences or interferences. Should one, then, disallow all forms of professional medical judgments as the basis for 'explanation' or 'justification' of social behavior interferences? How does one distinguish between an organically determined weakness or inability to work and a poorly motivated person with a non-psychiatric disease who is unwilling to work? Many of the criticisms Szasz singles out and directs at psychiatry and psychiatrists, to repeat, can be applied to all medical disciplines, and criticisms about how physicians handle psychiatric diseases apply to the way other diseases are handled.

Here I am pointing out that, if one chooses to view disease from a different, more general frame of reference, then many of the problems involving psychiatry become less striking, indeed. The papers of Toulmin, Feldstein, van den Berg, and my own all seem to direct to this point. The basic dilemma, to me, seems to be: biomedicine has 'made' disease a non-behavioral and non-person thing. (Recall the message of van den Berg.) Many want disease to be a non-behavioral thing and to stay there! If disease is handled in this way, then how do we in fact handle the notion of illness – a social-behavioral interruption – in a prudent way? How should we resolve social crises involving illness and who will do so? How will we link the taxonomic language of biomedicine with the language of illness? Szasz's views may be interpreted as warnings inherent in the unchecked course and direction of biomedicine. In my views, the logic of the biomedical taxonomy seems to lead to the separation of the idea of illness from that of disease. This separation contributes to many of the problems posed by psychiatric disease.

that so-called mental disorders are successfully assimilated to the concepts of health and disease, to that extent particular judgments of mental illness count as findings. One may be led to conclude that the model of health, in the context of psychiatry, is a mixed model that shows clear affinities with the models that prevail in general medicine, and, at the same time, with the models of happiness and well-being that prevail in the ethical realm.

The challenge is to formulate a single, comprehensive model for medicine, which does not exclude psychiatry. Such a task would not be acceptable to those who attack psychiatry for its supposedly blind allegiance to the medical model and believe that some of psychiatry's functions belong outside of the medical model. These critics recommend the removal of the functions now performed by psychiatry from the conceptual and professional jurisdiction of medicine and their reallocation to a new behavioral science-based discipline. Medicine then would be responsible for the treatment of disease while the new discipline would concern itself with re-education of people with their 'problems of living.' Suggested in this argument is the premise that while the 'medical model' constitutes a sound framework within which to understand and treat disease, it is not relevant to the problems classically believed the domain of psychiatry. In other words, psychiatry's chief problem, according to such an argument, is its failure to appreciate the inappropriateness of the 'medical model' for its tasks.

The medical model is usually thought of in terms of the biomedical, assuming disease to be fully accounted for by deviations from the norm of measurable biological variables. Such a narrow model or concept of disease is quite inadequate, and always has been, for psychiatry or any other area of medicine. Toulmin, in this symposium, calls this narrow approach a mechanistic model that sees all medical complaints as springing from 'mechanical faults' in the patient's body [11]. A biopsychosocial model is essential for understanding the determinants of disease and arriving at rational treatments and patterns of health care. As George E. Engel succinctly states: 'By evaluating all factors contributing to both illness and patienthood rather than giving primacy to biological factors alone, the biopsychosocial model would make it possible to account for the fact that some individuals will experience as illness conditions which other individuals regard merely as "problems of living" [3]. Engel goes on to emphasize that the scope of the biopsychosocial model is determined by the historic function of the physician to establish whether one is 'sick' or 'well'; and if sick, how sick, why sick, and in what ways sick; and then to develop a program to treat the illness, and restore and maintain health. And included in the physician's function or role is that of educator and psychotherapist.

Early in this century Sigmund Freud, Adolph Meyer, and their likeminded colleagues, re-emphasized a recurring approach in medicine's history of bringing together biomedical and psychosocial developments, but their efforts were overshadowed by the triumphs of biomedical advances and the lack of development in the behavioral sciences. As a result, medicine began to move away from the total patient (or whole person with a disease) and to concentrate on the pathology or the disease process itself. The pendulum is now swinging back and medicine is searching for a more adequate model – a blending of the psychodynamic and the reaction to life stress approaches.

In the light of the blending of these approaches, how then does the physician begin work with the patient in his effort to understand what is happening to him. He recognizes that illness, physical or mental, is a meaningful phenomenon, that its language can be understood and, therefore, should be listened to. Much has been written about the language of the body and how both patient and physician should listen to what the body is trying to say through its illness. Fabrega, in this symposium [5], and in a number of other important writings [4, 6], illuminates this issue in his discussions of the taxonomy of disease as a special type of multileveled 'cognized' model.

Present knowledge of any disease category makes the physician cautious about identifying any one factor as the cause of illness. Fortunately, medicine is abandoning the rigid cause and effect determinism which has recently dominated the diagnosis and treatment of illness and is thinking more in terms of multiple causation, critical phases of development of the individual, and the more universal factors which play a part in both health and illness. Thus, it seems essential to speak of a constellation of factors which enters into the development of a particular illness, with each factor perhaps weighted differently. The constitution of the individual, his social setting, the role of life experience, his inherited as well as his acquired immunity, and his physiologic and psychologic make-up are all factors related to causality. It is well accepted that emotional determinants can represent the *critical* cause in the appearance and remission of structurally manifested disability and pathology. Surely questions which focus sharply on what is organic and what is functional have no real meaning in present-day medicine. Man is a product of his total environment and disease is a protest of the entire person.

That no approach be neglected, that we become neither prisoners of anatomy, nor the anatomy of thought, let us embrace the words of wisdom from a Greek philosopher – medical wisdom, which will last as long as people will care for people. A young man, Charmides, in one of Plato's dialogues, complains about a headache. He wants a particular drug, but Socrates explains to him at length that this simple treatment is not adequate. 'To treat the head by itself, apart from the body as a whole,' he says, 'is utter folly.' The ideal approach had been described to him (Socrates) by a Thracian physician:

> You ought not to attempt to cure eyes
> Without head,
> Or head without body,
> So you should not treat body
> Without soul [9].

It seems indicated at this point to ask about the implications in treatment or recovery when the comprehensive medical model is used in psychiatry. For example, since the 'causes' of mental illness may include self-persuasion and self-delusion, the 'treatment' of these illnesses may involve a reversal of the precipitating factors – that is, counter-persuasion. The alternative (as Thomas Szasz has chosen) is to deny the acceptability of speaking at all of mental illness as illness. Yet, once the affinities and the differences have been pointed out, it does not matter whether the label is withheld or granted, for one understands the sense in which mental illness is held to be an illness and the sense in which it is not. The crucial point in such cases is the loss of the capacity of choice and freedom. Since the paradigms of free and responsible behavior are to be found in the moral and legal domains, it is not surprising that psychiatric treatment should at times resemble, or imitate, ethical judgment and advice. In so doing, it will inevitably generate some jurisdictional quarrels about values. Thomas Szasz has made a major contribution to the mental health field in focusing for over two decades on the issues and their implications in these jurisdictional quarrels. He has taken on his own specialty of psychiatry in calling attention to the looseness of its language, its diagnostic inaccuracies, the potential abuse of its power, as well as a multitude of other problems and pretensions of this speciality. While accepting a great deal of what Szasz has proclaimed, there is ample room to disagree with him sharply on many of his contentions. 'So long as a man is trying to tell the truth,' wrote John Jay Chapman, 'his remarks will contain a margin which other people will regard as mystifying and irritating exaggeration. It is this very margin of controversy that does the work . . . . No explosion follows a lie' [1]. It seems that Chapman has characterized well Szasz's work. There are many places in his paper where sweeping statements are made or crucial material appears omitted from his argument [10]. Just the same, he challenges us to wrestle with the issues he raises and confirms the relevance and necessity of symposia such as this one.

Another live issue in mental health is freedom versus determinism. One should never forget that in spite of Freud's deterministic viewpoint and his great emphasis on the past as the conditioner of the future, he involved himself in the treatment of patients with the intention of liberating them, at least in part, from the bondage of the past. Thus, no matter what might have been his theories, in practice he accepted man's capacity to change, thereby acknowledging the existence of freedom in man. If one fails to recognize this freedom, he could surely not in good faith participate in psychotherapy either as therapist or as patient.

There should be some way to look at this issue of determinism and indeterminism in a way which would be acceptable to the groups who view the issue in an opposing manner. Possibly freedom can be defined as a human capacity for choosing among possibilities of achievement combined with the capacity for action related to the choosing. Such an interpretation of freedom really does not conflict with theories of determinism which stress the role of the past in human behavior. From the standpoint of human behavior, determinism usually looks backward for retrospective analysis, whereas freedom looks ahead for a prospective evaluation of human achievement. It is interesting that Aquinas and Dewey have both suggested that freedom is related to self-determinism. This interpretation of freedom does not imply some kind of miracle which is subject to no law, for freedom has its limitations and its order. Much of the difficulty in resolving the determinism–indeterminism issue is related to a tendency to think of events as necessarily happening to something or to someone, and a tendency to separate past, present, and future in our conceptualizing. Any psychological moment contains within itself the past, the present, and the future. No particular moment can be frozen, for in the midst of an ongoing process the established form of the past meets the undetermined of the present in a continually emerging creative synthesis.

It is interesting to note that St. Augustine bypassed the conventional form of the problem of 'freedom versus determinism' with his extraordinary notion of the different orders of time: past and future time on the one hand, and present time, on the other [7]. The past comprises the processes of recollection (*memoria*) and the future the processes of prediction (*exspectatio*). St. Augustine emphasized that present time, however, is a variable 'moment,' which may be distended or shrunk. He called this 'contuition' (*contuitus*). Past events are now determined and future events are even now determinate. The present moment does not fall within the order of clock time and is thus outside of the causal order. Thus, it is experienced not as sequence but as spontaneity.

Alan Donagan, in this symposium, has stated that the essence of mental disease be identified, not with any homeostatic malfunction, but with loss of control of actions that are normally voluntary, or loss of normal power to arrive at the truth about oneself or one's situation, caused by the pressure of repressed affective energy [2]. In accepting such a view, one sees the validity of psychiatry's claim that a major mission is to increase human freedom.

Probably the mental health field faces more ethical dilemmas than other areas of health. Many of these dilemmas involve balancing the needs and rights of the individual against those of the community – the deontologic versus the utilitarian. Among these ethical dilemmas are the conditions under which civil commitment is warranted; the patient's right to receive treatment or to decline it; when, if ever, to use psychosurgery to treat what may be classified as a functional disorder; how scarce resources related to health are to be distributed without being discriminatory; the patient's right to be informed about his condition; the confidentiality of patient communications; and the issues of informed consent and risk/benefit ratios in human experimentation. Many of these, of course, refer to the health field in general and not just mental health.

Along with the ethical issues are many others attracting attention and controversy that relate to etiology, diagnosis, and treatment. Some are of a more general nature, such as the recognition of ethical harmony with one's fellowmen as a substantial basis of mental health. Another issue attracting attention is the acknowledgment that psychosexual development is a lifetime process, with new operative input taking place throughout life. The recognition of the influences, for example, that operate psychosexually from within and without on the geriatric age group could well change our values about aging, and our understanding of the final stage in human development. The bringing of prevention and public health into health care delivery systems while avoiding the pitfalls of paternalism or infringement on individual or collective rights offers another challenge.

These issues are being debated by many groups within and without medicine. Out of debate and controversy will come, it is hoped, insights and intellectual growth that will help us examine our life and work. In all of our endeavors, may we have the openness expressed by Louis Pasteur in *The Germ Theory and Its Application to Medicine and Surgery:*

> I desire judgment and criticism upon all my contributions. Little tolerant of frivolous or prejudiced contradiction, contemptuous of that ignorant criticism which doubts on principle, I welcome with open arms the militant attack which has a method in doubting and whose rule of conduct has the motto 'More light' [8].

Texas A & M University,
College Station, Texas

BIBLIOGRAPHY

1. Chapman, J. J.: 1900, *Practical Agitation*, Scribner's, New York, p. 51.
2. Donagan, A.: 1977, 'How Much Neurosis Should We Bear?', in this volume, pp. 41–53.
3. Engel, G. E.: 'The Need for a New Medical Model. A Challenge for Medicine and Psychiatry', unpublished manuscript.
4. Fabrega, H., Jr.: 1974, *Disease and Social Behavior. An Interdisciplinary Perspective*, MIT Press, Cambridge, Massachusetts.
5. Fabrega, H., Jr.: 1977, 'Disease Viewed as a Symbolic Category', in this volume, pp. 79–106.
6. Fabrega, H., Jr.: 1975, 'The Position of Psychiatry in the Understanding of Human Disease', *Archives of General Psychiatry* **32**, 1500–1512.
7. Outler, A. C.: 1963, 'Anxiety and Grace, an Augustinian Perspective', in S. Hiltner and K. Menninger (eds.), *Constructive Aspects of Anxiety*, Abingdon Press, New York, pp. 97–98.
8. Pasteur, L.: 1968, *The Germ Theory and Its Application to Medicine and Surgery*, in M. B. Strauss (ed.), *Familiar Medical Quotations*, Little, Brown and Company, Boston.
9. Plato, *Charmides*, 155–156/LB 15–21.
10. Szasz, T. S.: 1977, 'The Concept of Mental Illness: Explanation or Justification?', in this volume, pp. 235–250.
11. Toulmin, S.: 1977, 'Psychic Health, Mental Clarity, Self-Knowledge and Other Virtues', in this volume, pp. 55–70.

KAREN LEBACQZ

A man declares that he is Jesus. What shall we make of it? First, we might take it for just what it is: a person asserting that he is the Savior. Second, we might respond to it plainly, matching it against our own knowledge of the world: a person asserting a false identity; in short, a liar. Third, we might respond to it psychiatrically, matching it against our knowledge of psychiatry; a person displaying a delusion; in short, a psychotic. (T. Szasz, [4], pp. 243–244).

The scenario posed here is not much removed from an actual incident in my life. Several years ago, a conversation partner suddenly shouted at me: 'You goddam nurses can't hold me here in this Korean hell-hole forever; I'm damn well going to get myself a little cunt.'[1] There followed a series of curses and threats against the government.

Since I am not a nurse and was not in Korea at the time, I found myself in a situation analogous to Szasz's scenario, though perhaps a bit more frightening (I was the only 'cunt' in sight). How should I respond? Should I comment on the presumed war in Korea, although the year was 1972 and not 1952? Should I call him a liar? Should I run away? Should I call for help? – and if so, to whom? The police? The mental authorities? Family? Friends?

This experience haunted me, and I read the essays for this symposium eagerly in hopes that they would tell me something about this situation –

how to respond to aberrant behavior and when to judge someone 'mentally ill.'

Unfortunately, I got little help from any of the essays submitted here. Szasz argues that to label this 'mentally ill' is to judge and not to describe him. I have no quarrel with this claim; I simply find it unhelpful. Every response to a situation depends upon an interpretation of that situation and every interpretation is in some sense a 'labeling.' When I respond to my colleagues on this panel by taking their statements seriously and replying with rational discourse, I have 'labeled' those colleagues just as much as if I took their statements to be signs of 'mental illness' and called the mental health authorities. My perception of every situation I encounter is mediated by a framework of interpretation that has both cultural and personal elements; nor is there any way to avoid this. Thus we cannot achieve 'truth' by avoiding interpretation and labeling, as Szasz appears to wish. He may be right that the psychiatric labels are no more 'true' than other interpretations of the situation; but what he fails to give us is any index for choosing the more adequate or 'right' interpretation.

Toulmin acknowledges this central question when he proposes replacing the 'disease-cure' framework with a 'complaint-remedy' framework. But he stops short of telling us when 'complaints' should be considered 'diseases' and how to decide what 'remedy' to bring to bear for the complaint confronting us. The one criterion offered is that of the 'virtue' of alternative available remedies. In contrast to Szasz, Toulmin finds at least *some* virtue in psychiatric remedies.

In like manner, Fabrega suggests that every interpretation/diagnosis/labeling process by which we define an entity brings with it a prescription for control/elimination/response and that the accuracy of the diagnostic category chosen depends to some extent on the 'success' of the response. But what is 'success?' Am I successful if I avoid getting raped? – I return the conversation to the year 1972 instead of 1952? The first problem with the approach that looks to 'virtues' or 'success' to determine the adequacy of a label or response is the definition of success.

But there is another difficulty here. Many of us instinctively feel that the 'rightness' of a response depends not simply on its 'success' or consequences, but on its 'fittingness' to the initial situation. Suppose we can agree on what constitutes a successful outcome; still there might be several means to achieve that end, and the choice between those means depends on something besides the net effect. Listening quietly or injecting a drug may both have the effect of calming someone down; how do we choose? Suppose that the labeling of someone as 'mentally ill' had *good* effects (and not the bad effects that Szasz protests); would that fact make it right to label someone mentally ill at random?

Clearly, many of us would say No, and we would look to some aspect of the initial situation to tell us whether this label was appropriate. That is, we would argue that the label must relate to something in the actor or the initial action, something that 'justifies' the application of this label as over against some other label. Several symposium essays take this approach.

A popular form of this approach distinguishes 'mental health' from 'mental illness' by the 'freedom' or 'rationality' of the actor. Thalberg's discussion of free will and coercion demonstrates amply the problems with such an attempt.[2]

Donagan proposes the attribution of 'mental illness' where the person's actions are *inappropriate*: 'In themselves, the actions of neurotics and psychotics are like those of anybody else, but they are inappropriate to their situation' ([2]. p. 41). Thus the cursing related above might have been approprate in a hospital during the Korean War, but was not appropriate on the street in 1972. But note that the curses are inappropriate only from the perspective of *my* perceptions of the situation: if, when this man looked at me, he really saw a nurse in the Korean War keeping him confined, then his responses were not inappropriate to *his* perceptions. It is then the perceptions themselves that are at stake. Thus I agree with Thalberg when he says that 'peculiar beliefs, even beliefs about who one is, often dominate cases where we feel that the person has lost command of (*sic*) behavior' ([5], p. 211), and that it is the beliefs that seem out of whack.

'Mental illness,' then, may rest not so much in actions as in perceptions. Donagan suggests that there are two aspects of distorted perceptions in mental illness: first, 'loss of normal power to arrive at the truth about oneself or one's situation,' and second, that this loss of power is 'caused by the pressure of repressed affective energy' ([2], p. 51). This definition seems to relate well to the case at hand; my interlocutor appears to have lost his normal power to perceive that he is in 1972 and not in 1952, and a layperson would certainly judge that there was 'repressed affective energy' (both sexual and aggressive) involved. So perhaps I would be justified in choosing the label 'mental illness' for this situation and responding accordingly.

But, wait: I am a feminist. That is to say, it is my perception that women in this country have been and are systematically and institutionally discriminated against – as evidenced, for example, by the pervasive use of masculocentric language at this conference. (Any visitor from Mars would surely think that all physicians, all patients, all philosophers, and indeed all human beings were male!) This definition of the situation has been challenged, and some would say that I have lost my power to arrive at the 'truth' about my situation. Indeed, some have gone so far as to claim that the feminist perceptions derive from 'repressed affective energy.' In the vernacular, this theory is known as 'all she needs is a good lay' and is not an uncommon response to a feminist claim. Nor is the theory confined to the vernacular: a recent article in the *Journal of Psychiatry* argues that some feminist students use the women's movement as a place to consolidate their sexual identity and that with therapy the movement will no longer serve 'as a cognitive and affecting organizing focus around which sexual behavior pivots.'[3] By Donagan's definition, then, at least some would judge my feminism to be a form of 'neurosis.' As it is, I can only say it is one form of neurosis of which we need more, not less!

But the question is, Who is to judge that my feminist perceptions (or the perceptions of my interlocutor) are (a) wrong – i.e., not the 'truth' in the situation, and (b) the result of 'repressed affective energy?' The addition of the second qualification here – that the perceptions must be due to certain phenomena – seems to require a 'professional' judgment. Physicians (psychiatrists) decide who exhibits 'repressed affective energy,' and thus physicians must decide who is 'mentally ill.'[4] It seems, then, that I ought not to have decided *myself* that my interlocutor was 'mentally ill' but ought to have called the mental health authorities.

Yet this is troubling. Physicians *may* be competent to determine the presence of 'repressed affective energy,'[5] but it is not at all clear that they are competent to decide whose perceptions are wrong. While they might appropriately judge the presence of the second requirement, therefore, they may not appropriately judge the presence of the first. 'In the last analysis.' as Fabrega puts it, 'it is the social group' that must make the judgment as to what perceptions are wrong ([3]. p. 98).

Yet here we are faced with a 'Catch 22' situation. The social group must judge the rightness and wrongness of perceptions. Yet our social group turns that judgment over to various professional groups! Worse yet, it provides no central clearinghouse to tell the troubled individual *which* professional group to choose in a crisis of perception. With no central clearinghouse, I must make the decision myself as to how to interpret and respond to this case. As Toulmin notes, the professional specialization that we have today 'is liable to bear down very hard on the people who really need help' ([6], p. 68) and, I would add, on those of us who try to help them.

So what did I do? And how did I choose that response from among the many available alternatives?

First, I recognized that what was at stake was a question of perceptions of the situation, and that no matter what interpretive framework I chose there might be deleterious consequences (ranging from rape of me to the long-term incarceration of the other person). Thus, I decided to seek an alternative interpretation of the situation. I chose not to seek that interpretation from those who would have ready-made psychiatric labels, but from those who had known this man longer than I and for whom his actions might not be as aberrant as they seemed to me. Thus I hoped to minimize the chances that he would be brought to task under the rules of *my* particular social group. I called a friend of his.

Whether that was the 'right' thing to do I leave for others to decide. But that is not the end of the story. His friend called his family. They called the police. And the police called the mental health authorities, who took him away for 'observation' and subsequently declared him 'mentally ill.' I wonder what we would have done if he had only said he was Jesus Christ . . . .

Pacific School of Religion,
Berkeley, California

NOTES

[1] The language is reproduced as accurately as possible; other details have been altered slightly for purposes of conceptual clarity.

[2] Perhaps in a similar vein, Feldstein proposes that a 'healthy' person may be distinguished from an 'unhealthy' one by the 'luminosity' of the person's unconscious. I confess to finding his poetic description unintelligible, and I therefore pass on to other suggestions.

[3] Defries [1]. In fairness, it must be noted that the author does not claim this phenomenon to be true for *all* feminist students but only for *some*.

[4] The monopoly of definition held by physicians is described by both Szasz and Fabrega.

[5] I say 'may' because this determination itself involves a labeling and hence a valuing process.

BIBLIOGRAPHY

1. Defries, Zira: 1976, 'Pseudohomosexuality in Feminist Students', *American Journal of Psychiatry* **133**, 400.
2. Donagan, A.: 1977, 'How Much Neurosis Should We Bear?', in this volume, pp. 41–53.

3. Fabrega, H.: 1977, 'Disease Viewed as a Symbolic Category', in this volume, pp. 79–106.
4. Szasz, T.: 1977, 'The Concept of Mental Illness: Explanation or Justification?', in this volume, pp. 235–250.
5. Thalberg, I: 1977, 'Motivational Disturbances and Free Will', in this volume, pp. 201–220.
6. Toulmin, S.: 1977, 'Psychic Health, Mental Clarity, Self-Knowledge and Other Virtues', in this volume, pp. 55–70.

THOMAS SZASZ

I said what I planned to say in my formal paper, and am unable to offer any specific remarks on the many papers and discussions that were presented at this conference. I should like to confine myself to a single observation that struck my attention concerning the conference as a whole.

I believe that the meaning of a word, especially in psychiatry and politics, must be sought in the practices it implies; in other words, in its consequences. The consequences of 'having' a medical illness or a mental illness – or 'having' diebetes or schizophrenia – are very different. Yet, most of the speakers approached 'mental illness' as if it were, *prima facie*, similar to medical illness; as if 'mental illness' meant what psychiatrists, or philosophers, say it means. As a result, abstract, scholarly distinctions between various views of 'mental illness' were illuminated, while the actual conduct of 'mental patients' and psychiatrists, and the moral and political implications of their acts, were obscured.

For me, individual liberty is more important – that is, more precious morally – than 'mental health,' no matter how 'mental health' may be defined. Hence I continue to be wary of 'scholarly' approaches to 'mental health' lest, by legitimizing this concept, they contribute, however unwittingly, to the traditional psychiatric infringements of human dignity and liberty.

BERNARD TOWERS

This is not an easy collection of papers to review, because they are so diverse in both form and content. Philosophers tend, more than most critics, to be idiosyncratic in their approaches, and the theme 'mental health' has given maximum opportunity for individualistic treatment. There is no area of modern medicine that is more fractured, as regards both theory and practice, than is psychiatry, the study of the mind in both health and disease. We have the major groups such as advocates of psychoanalysis and no drugs; and growing numbers of advocates of the reverse; advocates of behavior-modification techniques as the only treatment of the psyche that has predictable and guaranteed results; and advocates of an eclecticism that would allow for a bit of this and a bit of that, with little hope of agreement from any other therapist. Compared to the study of infectious diseases, or congenital heart disease, psychiatry appears to be in total disarray.

It is not just the major groups outlined above who fail to find much common ground. Within the ranks of 'pure' psychiatry, based on analysis of the psyche and the influences both conscious and unconscious to which it has been subjected throughout one's lifetime, disagreements seem to be as much the order of today as they were in Vienna soon after it all got started. In Southern California it looks as though one, at least, of the venerable Psychoanalytic Institutes is in danger of being riven apart by the disagreements between the Freudians and their Kleinian rivals. Some measure of disagreement in both theory and practice is, of course, a sign of health in any discipline. The last thing one wants to see in the study of the mind (that most complex manifestation of the power of the process of biological evolution) is a rigid orthodoxy that will admit of nothing new. And yet, somewhere between a too rigid orthodoxy and a too flexible anarchy there must be a system of thought that embraces the best elements of each.

It is a very interesting fact that at a time when not a few psychoanalysts seem to be losing faith in their own achievements, professional philosophers seem to be entering the field (as at this conference) with marked success. I am thinking here of the recent collection of critical essays edited under the title *Freud* by Richard Wollheim in the *Modern Studies in Philosophy* series [3] and of other essays such as Wollheim's 'The Mind and the Mind's Image of Itself' [4]. It may be that there is now a sufficient corpus of writings in Freudian Metapsychology that will make it possible for metaphysicians to do for the Unconscious what they have done for so long with respect to the Conscious Mind.

But there is a major drawback to the employment of the philosopher's *ratio* (reason) – that finely-honed instrument of logic – for the development of the new discipline of metapsychoanalysis. It is this drawback, and the failure to recognize it (or at least to recognize it publicly) by the contributors to this conference to which I now want to turn. One can read about and talk about the *concept* of the Unconscious with impunity. Words are cheap, and words about words even cheaper. But no amount of reading or talking can give one any experience of or real insight into the Unconscious itself. The only place where understanding can properly be sought is the analyst's couch. It takes many long hours, much free association, and the recounting of many dreams before the often bizarre workings of the Unconscious become in any way intelligible. And even when one does begin dimly to discern what goes on in those more impenetrable regions of the mind, and even more what *must* have gone on in times long-past that caused one to repress the thoughts and fears, the expectations and one's regrets at their non-fulfillment – even then it becomes difficult if not impossible to express one's conclusions in language that can be understood by anyone other than oneself and one's analyst.

Since, as stated earlier, there is such a plethora of psychoanalytic and other psychiatric theories from which to choose, and since one's analyst (or a succession of them) has such a powerful influence over one's own interpretations, then clearly it behooves a philosopher or a physician writing on these themes, to indicate in some way where he comes from in his search for understanding. This is not to insist on a paper's being prefaced by a psychobiography of oneself. Rather it is to suggest that one might at least indicate whether or not one has had any experience of depth-analysis, and, if so, what *kind* of analysis (whether Freudian, Jungian, Kleinian or whatever) and, more particularly, whether one thinks it was a worthwhile and 'successful' experience, or one that one would rather forget or deny. If it is the case, as Alan Donagan argues in his paper (and I for one believe it to be true), that 'everybody is at least incipiently neurotic,' and that acquisition of knowledge of one's own neurotic traits might either remain at the 'merely theoretical' level or else be made truly 'effective,' as Donagan puts it, by the conscientious working through, in analysis, of formerly repressed conflicts ([1], pp. 48, 51), then the analyst's role as guide and teacher will be of crucial significance. It is true, of course, that insights into the Unconscious cannot be 'wished' upon one, or imposed from outside. The analytic process, when properly undertaken, is a process towards greater self-understanding, towards greater autonomy, and hence towards greater human freedom of action. But if improperly or inadequately conducted, the process can clearly work in the opposite direction, and one can end up much less free from the constraints of the Unconscious, and much less able to view oneself, one's analyst, and the process itself at once objectively and yet with that degree of empathy which is a prerequisite of all human understanding.

It has been facetiously said that the most important choice we ever make is the choice of whom to have as parents. Next in importance is surely the choice of analyst, that proxy-parent who tries to see us safely through the reliving of our pre-verbal conflicts, stresses, and denials. It is generally accepted (much to the chagrin of the highly-committed) that it does not make a great deal of difference to which theoretical school the analyst belongs, provided that it is a school wherein the patient is encouraged to grow and develop, freed from the initial dependent role. It will be important, if the 20th century is to be accurately assessed so far as the psychoanalytic movement is concerned, that good thinkers, independent logicians, should help us sort out the wheat from the chaff in analytic theory. But what seems to me to matter most with regard to the study of mental health and mental disease, is that philosophers and other non-medical thinkers should themselves undertake genuine psychoanalytic experience, whether as patients or as training analysts or (if the financial outlay were not too big for most of us) simply as students anxious to increase our understanding of ourselves and others. If a philosopher tells me he has never experienced 'guilt,' I know at once where he stands – in abysmal ignorance of the human condition. Again, when someone talks or writes about 'love,' he may be skillful enough with words to hide the fact that he has never actually experienced love. And yet words about 'love,' like words about the 'Unconscious' are 'as sounding brass or tinkling cymbal' if they are based merely on vicarious experience.

How is one to insure that the study has been properly conducted? The academic world has a well-constructed review-system that can more-or-less guarantee the level of training that an individual has received. The system has its faults, to be sure, but when one reads a list of symposium faculty such as there is on the progam for this conference, then one can be fairly sure that the degrees listed and the current academic positions held are proof of the individual's acceptance by academic peers. How can one be sure about opinions in the field of the psyche if one does not, is not permitted to, know in which milieu those opinions germinated?

In the absence of any such information (other than what can be construed from internal evidence) I suggest that speakers at conferences such as this be encouraged to give brief statements concerning their qualifications (other than command of vocabulary and a measure of articulateness) to speak about the psyche. This might be done primarily by reference to their particular orientation, or to the school from which their mentors came. Orientation and quality of the therapist is more important that his or her identity; and after all, the confidential nature of the analyst-analysand relationship goes both ways. But openness about the kind of experiences one has, or has not, undertaken, is surely vital to assessment of the worthwhileness of such value-judgements as may emerge in a theoretical paper on Mental Health.

For myself, my professional studies of human biological development both in ontogeny and phylogeny led me naturally to many years of reading psychoanalytic and other psychological theories of the mind's development, but all at the 'merely theoretical' level to which Donagan refers. Some fifteen years ago I began to seek more 'effective' knowledge, and for some years in England I sought help, primarily through dream-analysis, from two eclectic practitioners and also from a Jungian analyst. Emigration to America and the strains of adjustment to my new milieu exacerbated my anxieties about who I was or was to become. For a year I did regular sessions with a Freudian who had a philosophical bent that was to my liking. But still I was operating only at the theoretical level, and found that the sessions merely allowed me to deny and escape from myself through constant *talking about* the Unconscious instead of getting to *know it directly*. So I changed analysts again. I have now completed three years of what is often really hard work under the guidance of a very skillful and very orthodox Freudian analyst. Sometime before the present decade is over I hope to bring my present inquiry to a successful termination in the formal sense. There clearly will be no end to one's personal analysis of one's own psychohistory. Illumination, and the emotional stability that it affords, come slowly and painfully, with many set-backs that serve to induce a fitting humility about one's minor achievements in this difficult field.

I am sure that some of the speakers and other participants in this symposium will want to say that I am wasting too much time and effort and money in pursuit of enlightenment by this kind of experience; or that by deliberately casting myself in a patient-role, a sick-role, I am defeating the very object of the exercise. I can only reply that for me my current experiences are the only ones that have ever made real sense out of that exploration of those other dimensions of thoughts and feelings that are beyond the range of the philosopher's reason, that are beyond our daytime logic. Some psychoanalysts are now pursuing Erikson's well-accepted stages of psychic growth in children into further stages in adult life. The Goulds, in their recent paper on 'The Relationship Between an Individual Growth Step and Marital Discord' [2], seem to imply that mental disease or distress is a normal and even essential ingredient to psychic growth according to the psychoanalytic model. I think that is true.

Some adopt other and maybe more successful modes of inquiry. I cannot tell from internal evidence the route by which Dr. Feldstein has travelled towards the vision of unity that he expresses in his paper with such powerful poetic imagery. His language seems at times to verge on the mystical. Maybe someday, after all my years of analytical plodding through dreams and fantasies, and the forgotten feeling-memories of infancy and childhood that they sometimes evoke, I too will be granted some intimations of that inner unity which Feldstein expresses as the meaning of 'symbol.' His vision strikes me as being very Teilhardian in character. For me it has great appeal. But there is much hard work to be done before it will be generally palatable in philosophical circles such as this.

Center for the Health Sciences,
University of California at Los Angeles,
Los Angeles, California.

REFERENCES

1. Donagan, A.: 1976, 'How Much Neurosis Should We Bear', in this volume, pp. 41–53.
2. Gould, R. L. and R. V.: 1976, 'The Relationship Between an Individual Growth Step and Marital Discord', paper read to the American Psychiatric Association May 11, 1976.
3. Wollheim, R. (ed.): 1974, *Freud: A Collection of Critical Essays,* Doubleday, New York.
4. Wollheim, R.: 1969, 'The Mind and the Mind's Image of Itself', *International Journal of Psychoanalysis* **50**, 209–220.

EDMUND L. ERDE

The semantics of 'health' seems to have at least three facets: an evaluation aspect, a prescriptive force, and a descriptive content. These depend on tacit norms which may vary with the context (in terms, say, of the age of the person being considered, or the reason for inquiring into his or her health). Nevertheless, when pondered in the abstract fashion that this conference encouraged, tensions may be expected among theorists who believe in mental health over and against physical health (as Freudians do), those who do not believe in that distinction or the mind–body distinction which seems to foster that first distinction (Toulmin and van den Berg), and still other thinkers who do not believe that there is such a thing as mental health at all – that the idiom is misplaced somehow (Szasz?). They war, I believe, because each of the three facets of the concept *health* calls for different analyses and responses. Trouble arises when an insight or discovery about a difficulty concerning one facet is used to assault the whole concept. It is, for example, no doubt the case that mental health terms (intentionally or not) are abusively employed for gain or motives other than care of the afflicted. In opposing such abuse one might for rhetorical considerations or through confusions deny 'health' and other terms their appropriate descriptive import. This may even be half-right, for certainly how we employ a concept in evaluating and prescribing is part of its meaning.

Many contributors to the symposium have made us more aware of the abuses of 'health,' etc. Unfortunately no one has parried by raising our awareness of the constructive point behind the concept *mental health* so that it could be thus abused. For example, mental illness *saves* people from certain kinds of punishment or deprivation. Some people find this bad, but sometimes it is good. For instance, the English sovereign was once able to deprive a family of an estate if the leader of the household was 'guilty' of suicide. Attributing insanity to the act was a way of helping the family to keep their proper due.

The abuses of the prescriptive and evaluative facets of 'health' warrant our attention and repair, but this does not mean that those facets of the grammar of 'mental health' do not otherwise play an important and constructive role. And it certainly does not mean that the descriptive element refers to the never-existent. In what follows, I try to supply some insights into the descriptive aspect -- an understanding of *mental health* and *mental illness* as these bear on the individual who suffers from loss of whatever is lost. Doing this might at least supply a warrant for retaining, in the treatment of the mentally ill, 'the pediatric model,' the parentalism which Dr. Szasz opposes. But such parentalism can only be licensed when the patient strays so far from the community's concept of *intelligence* and *rationality* that he or she resembles the very young child in important ways.

Not surprisingly several contributors to our symposium have considered mental illness immediately rather than mental health, as the symposium's title enjoins. Notable exceptions include Professors Toulmin and van den Berg, whose papers are especially exciting because they give the World Health Organization's notorious definition of 'health' a vitality and cash value which it has lacked heretofore. Their essays explicitly relate health (without adjectival modifiers) to every dimension of human life. Indeed, they each seem to reject the distinction between mental health and physical health and concentrate instead on *health simpliciter.*

Van den Berg shows how our ordinary distinctions between mental and physical health and illness have evolved over centuries of regrettable cultural–philosophical development. And he explains why I, living under the seventh milestone where neurosis is normal, in looking forward to this symposium, kept thinking that it was about mental illness![1] So I am going to follow the trend of illness (though I also long to attend to the notion of mental health in terms of *growth* and *maturation*).

When is mental illness attributed to an individual? It must be when and because his or her behavior appears irrational by community standards. And the concepts *rationality* and *irrationality* are most apparent by their absence from the discussion. Rationality and irrationality have to do with how understandable behavior is. Understanding behavior depends on presuppositions[2] about the beliefs of the agent who is acting. These beliefs are available to us primarily or exclusively through language.

Why is language so important? The heart of rationality is the ability to employ higher order canons in considering and communicating about the logical status of general claims about the world.[3] The ability to bring logic to bear on particular arguments characterizes rationality. Although the ability belongs to individuals, it is refined by cooperative interpersonal exchange. This kind of dialectic considers the role of ideals or categories essential to a paradigm of perception and cognition. Let me explain all this by considering a type of model of mental illness which permeates so many of the papers – mental illness as a cognitive or perceptual difficulty which I will call 'perceptual,' *meaning by it that cognition is built into perception.* This view of perception relates to gestalt psychologists and philosophers from Plato (in *The Theatetus*) through Kant to current analytic philosophy. Gestalt therapists have tried to explain the pathos of mental illness through such a model ([6], pp. 4–30).

First, perception is an activity. It is not the passive reception of sensory data which is later interpreted. Rather, it has built-in organization and is shaped by prior categories which make possible the parsing of the blooming buzzing manifold which would otherwise bombard us continuously. A conceptual framework is the logical basis for the possibility of perception. We could never perceive without the categorial filters which structure the seeing.

Consider, for example what is required to make the observation called 'seeing what the time is' ([9], chapter 1). One must know the conventions of reading the face of a clock, reading numbers and knowing what they are, knowing how hands point, and seeing where the hands are. And because 'to see' in its rich ordinary sense implies some accuracy, one has to know a bit about astronomy, geography, where one is, and how the clock works. Implicitly we rely on the shape of the earth, that it turns on its axis and revolves around the sun, as well as on the effect of latitude and seasons on daylight. All these are involved in reading the time, for that is not the same as reading a clock.

In accord with this, Thomas Kuhn [11] argues that science is not a cumulative activity but is essentially revolutionary. He calls conceptual frameworks 'paradigms.' A paradigm involves standards of observation, meanings of terms, classical experiments in the history of science, standards of questions allowable, and more. All these structure perception. Kuhn's view is that science progresses from a pre-paradigm state thence to a paradigm which would be articulated further by the people who accept it. As they work, they accumulate recalcitrant phenomena. Readjustments to these make the paradigm increasingly cumbersome. This may result in a crisis in which investigators normally bound by shared concepts lose much of what they share, losing then the sense of a community. Opposing camps often form with great fighting, for although (contrary to what Kuhn says) people within the same science who do not hold with the same paradigm do have some overlap of terminology, methodology, history, etc., their differences are more apparent than their similarities. When the crisis is resolved by either the victory of a new paradigm or reconfirmation of the incumbent paradigm, a reconstituted community resumes paradigm articulation.

People of R. D. Laing's view seem to believe that those whom we call mentally ill merely live in a different world from ours. To analogize from Kuhn, they live under a different paradigm. But perhaps such people should be seen as in a crisis phase, instead – in what J. L. Reed [14] in his study of schizophrenic thought disorders describes as a breakdown, not just a difference in the filtering of input. Reed discusses Broadbent [3], who specifies a physiological filter, and perhaps means it to be the locus or embodiment of what I am calling 'the paradigm.' If there were such a physiological mechanism and it were to be damaged, the perceptual bombardment would be horrendous and would likely precipitate some of the thought and speech disorders ascribed to schizophrenics. But (*pace* Broadbent?) an explanation of mental illness need not always be physiological. Such a breakdown might be due to trauma of the sort Freud discovered. A disruptive experience could fragment one's self-concept or give someone an idea so contrary to his existing paradigm about his world that it would put him into crisis. Hysterical blindness might be understood in this fashion. In this way, Kuhn's model might be more applicable to the mentally ill than it is to the growth of science.

Now, Freud seemed to have believed in the possibility of mapping components of the personality and psychic functioning onto areas of the brain ([8], vol. 2, pp. xxiv, 160, 193–207, 228–9, 285; vol. 23, pp. 144ff). But his clinical work was oriented toward helping the patients discover the ideas which had led them astray. This may have applied more to neurotics than, say, schizophrenics, but these subcategories of mental illness share an important feature. Professor Donagan's paper reminds us of it through the mechanical imagery invoked in psychoanalytic theory. He points out that nonbodily disease is probably drawn on an analogy to bodily disease, and he indicates how the notion of mental functions are originally copied from materialistic theories of functions. The mechanistic image is apparent in his language: e.g., (i) 'a straightforward product of the interaction of the various physical and psychical systems . . . . ' and (ii) the ego was constructed as 'one system among others' and 'figures as a quasi agent' ([5], pp. 46–47). The notions of *drive* and *desire* and the reification of the elements of the personality which interact according to conservation of energy are deterministic mechanical models.

Donagan is troubled by this imagery because it abandons the autonomy which marks the essential difference between persons and everything else. But Freud's point is compelling. He displays how much against the will mental dysfunction appears to be. One cannot avoid certain ways of thinking or acting upon certain tendencies. However wrong Freud might have been about the healthy personality, his insights draw attention to the causal feature proper to an account or description of the mentally ill individual.

Professor Thalberg, attentive to both the mechanistic and perceptual aspects of mental illness, makes explicit what is just presumed in many other papers: the warrant for the word 'ill' in 'mental illness' in terms of *suffering.* He reminds us that Freud focused on the generation and avoidance of anxiety, giving a schedule for distinguishing between neuroses and psychoses. If the anxiety is addressed by the ego's remaining true to the external world and repressing the id, neuroses result; if the id overcomes the ego and tears it away from reality, psychoses result.

The anxiety attendant upon the struggle between ego and id is also analogous to the crisis state of paradigm choice. The personality does not know which perceptual system to allow. It cannot tolerate their full-blown coexistence. So the crises stage has another place in the Freudian account of mental illness, e.g., at the crossroads between neuroses and psychoses. (Perhaps this should cause us pause, but I just mark it.)

A reconciliation between mechanical and perceptual accounts may be achieved by noticing that the perceptual framework structures in a mechanistic way, in that it does not merely give us organizational categories for experience, but *ipso facto* places constraints on experience as well. When constraints are not functioning vigorously enough we have the crisis or breakdown of the filter function or, what I suggest comes to the same thing, a breakdown in rationality.

Considerations involving rationality apply to perceiving particular events and to reconstituting the paradigm itself. Grammatical devices for making separate statements are crucial to it. Mental illness may now be thought of as irrationality – breakdown of the self. The self may be thought of as a story about both the world and the individual whose life it is. Events might force the breakdown of the narrative and force the person to withdraw from rationality. Then reasons become more like causes and a mechanical pattern reminiscent of conditioning might emerge.

Here there is a connection with the beetle in the box, the pet of some Wittgensteinians – private language.[4] The idea is that a radically isolated individual, in attempting dialectic with himself, would not have the *necessary* support and direction of a community in structuring his concepts. In falling away from community standards of behavior, thoughts, and feelings, in losing the distinctions between correct and incorrect, impression and correct impression, etc., the language loses all norms and thus loses direction and implication [12]. Thus the lack of connection between community and the individual (the most general community which specifies the paradigm) results in a breakdown of the filtering mechanism. This is why community norms at the conceptual level (as opposed to the level of opinion) are so important.

The world management which *a rational community* evolves includes (i) a stock of reasons for claims and (ii) means for assessing those reasons. Mental illness of perhaps all types is *an individual's* loss of community membership. It is the loss of standards of truth and confirmation and the resort to private standards of perception and expression. The mentally ill person will fail to mark as evidence what other people will mark as evidence. But, more importantly, the general rules of adequate evidence seem unavailable to him. The concepts, in the narrative oı mythos and ideology which we hold as a community, break down to the extent that not merely isolated beliefs or actions are alien, but the whole framework is. We might say that because the sense of the general is lost, the adaptability and creative novelty included in normal[5] living is also lost. There is not enough structure for rational life.

On the other hand, an individual must have a notion of *partial reason* as well as *conclusive reason.* Without the mentality to weigh evidence and frame tendencies to believe, reasons are closer to causes. There is *too much* structure for there to be rational life! Hence the more typical psychology of the empiricist camp (i.e., behaviorism) fails for normal people. In circumstances where an individual elects to respond to one of a wide variety of things or when in a variety of circumstances he or she responds to the same thing, the causal, mechanical model of action is inapplicable and the idea of *reasons-for-action* is appropriate. Through the uses of both diverse tenses[6] and universal judgments, contexts become so variable that behaviorism fails to explain or describe. This is because notions of selection and relevance which seem to violate the rigid logic of the behaviorist apply [4]. When flexibility of this sort is lost, the result squares closely with our impressions of some mental illness, e.g., schizophrenia. The afflicted cannot invest in the universal nor investigate it. They have lost the ability to relate to the abstract and have lost the ability to deny through their actions false characterizations about the general. Bridging large gaps in time and coordinating a large number of sensory inputs amount to breaking away from a stimulus–response model.

So although some filtering must take place, it cannot be too tight. There is a virtue in a certain looseness of thought, too. This is why we must have a tolerance for appreciating reasons as marks-in-favor-of-or-against-holding-and-acting-upon-certain-beliefs rather than as decisive ([15], p. 47).

Thus, language is important because 'only linguistic behavior can be appropriate or inappropriate to that which is not both particular and present' ([2], p. 87). A *system* of beliefs is framed and understood to be about the general, not the particular. Thus we may say that the irrationality of a particular belief is different from the irrationality of a paradigm. All mental illness may be closer to the latter. For the behavior that is most characteristically worrisome in connection with mental health is not the behavior of a lone act, as in the case of the irrational behavior of one father killing another because of a fight between their wives because of a fight between their children ([10], p. 1), but that in which there is persistent loss of those canons which are used to take that individual through community life.

We may characterize the rationality of our species by appealing to a Kantian notion of concepts as rules for the synthesis of experience. I am trying to characterize the kind of irrationality and, through that, rationality which applies within our species and bears on those whom we think mentally ill. In doing so I have developed the theme that mental illness is a perceptual disorder. It is the observer's view of the breakdown of the common view and thence the loss of free will in the behavior of an individual who belongs in that common view.[7] That is why it is so important to have the kind of help in delimiting the common view that efforts like Dr. Fabrega's supply. But it is also why dialectic is important as a therapeutic device. If there is a breakdown in the paradigm of the mentally ill individual, yet enough overlap with our own paradigm to allow for some dialectic, then forcing the ill to clarity might help them reconstitute the paradigm and rejoin the common view [1]. Thus, in more than one sense, they recollect themselves through the anamnesis of dialectic. Whether paradigm repair is achieved thus, or whether it is achieved by repair of some physiological mechanism, the 'illness' of '*mental* illness' and the 'health' of '*mental* health' can be understood on the perceptual, causal, rational analysis described in many of our papers.

University of Texas Medical Branch,
Galveston, Texas

NOTES

[1] This is not to suggest that I fully accept van den Berg's implications. The gist of his paper seems to be that mental illness is a concept with a history informed by the history of biomedicine and psychology, with dashes of the history of physics. That history has placed contemporary man in a spot which makes the lived-in world horrible and unnatural in some sense or other. His position is paradoxical, for if all Western man has suffered that history, why would some of us be 'normal' or 'well' and others of us not? Why would there be differences in the sicknesses the sick suffer? That is, why the specifics of the individual cases? Here Toulmin urges us to attend to particulars of cases and this must lead us to construct answers about individuals. But we must ask van den Berg for an account of the standard distinction between neuroses and psychoses, their subcategories, why they do not return given our needs, and how the history of our civilization informs individual cases. Is his conclusion that we are so wrong about the idiom 'mental health' or 'mental illness' that even the distinction between neuroses and psychoses and their subcategories must go? Other problems for van den Berg concern difficulties in understanding the apparently mentally ill in cultures other than our own. And how should we understand the sufferings and aberrations of individuals living well before the completion of the last milestone? Were the people who were being treated or abused for what we would translate as 'mental illness' differently ill when they lived in the era of the fifth milestone or in the era of the third milestone?

Professor Toulmin's paper leaves out the historical dimensions except in personal or individual terms. In this regard he is closer to our typical understanding of cases than is van den Berg. For, regarding cases, as Toulmin rightly points out, by just hearing a patient's complaint one could not tell whether the complaint was physiologically, socially, or legally based, etc. One could not tell which kind of remedy, if any, was called for. Toulmin's point, like van den Berg's thesis, is unhelpful regarding the task of clarifying the notion of mental illness, except insofar as it forces us to attend to why we would want such a concept. They pretend to reject the label and force us to rethink its job.

[2] Like reading the time. See discussion below.

[3] I am indebted to J. Bennett's *Rationality* [2] in much of this discussion.

[4] See Bennett, ([2], pp. 74–5), and Wittgenstein ([16], §293 and the range of topics listed under 'private' in the index).

[5] 'Normal' is Kuhn's term for paradigm articulation – living within a paradigm.
[6] Tense use is very complex: past, present, future, the perfect and subjunctive.
[7] This unfortunately is still inadequate, for it fails to parse cases of revolutionary creativity from cases of irrationality or sort the loss of free will (e.g., to a song that keeps running through one's head) from mental illness.

REFERENCES

1. Bandler, R., and Grinder, J.: 1975, *The Structure of Magic,* Science and Behavior Books, Palo Alto, California.
2. Bennett, J.: 1964, *Rationality,* Routledge and Kegan Paul, London.
3. Broadbent, D. E.: 1958, *Perception and Communication,* Oxford University Press, Oxford.
4. Chomsky, N.: 1959, rev. of *Verbal Behavior* by B. F. Skinner, viii, 478 pp., Appleton-Century-Crofts, New York, 1957, *Language* **35**, 26–58.
5. Donagan, A., 'How Much Neurosis Should We Bear?', in this volume, pp. 41–53.
6. Fagan, J. and Shepherd, I. L.: 1970, *Gestalt Therapy Now,* Harper and Row, New York.
7. Fingarette, H.: 1969, *Self-Deception,* Routledge and Kegan Paul, London.
8. Freud, S.: 1955, *The Complete Psychological Works of Sigmund Freud,* trans. by James Strachey, Vols. 2 and 23, Hogarth Press, London.
9. Hanson, N. R.: 1965, *Patterns of Discovery,* Cambridge University Press, London.
10. *Houston Chronicle,* Friday, April 16, 1976.
11. Kuhn, T. S.: 1962, *The Structure of Scientific Revolutions,* Vol. 2, No. 2, 2nd. ed., University of Chicago Press, Chicago.
12. Malcolm, N.: 1954, 'Wittgenstein's *Philosophical Investigations*', *The Philosophical Review* **63**, 530–559.
13. Murphy, J. M.: 1976, 'Psychiatric Labelling in Cross-Cultural Perspective', *Science* **191**, 1019–1028.
14. Reed, J. L.: 1970, 'Schizophrenic Thought Disorder: A Review and Hypothesis', *Comprehensive Psychiatry* **11**, 403–432.
15. Waismann, F.: 1968, *How I See Philosophy,* Macmillan, London.
16. Wittgenstein, L.: 1958, *Philosophical Investigations,* 3rd. ed. Macmillan, New York.

H. TRISTRAM ENGELHARDT, JR. AND STUART F. SPICKER

CLOSING REFLECTIONS

'Mental health' and 'mental illness' are not simply medical or psychological terms – they are social and political as well. The social status of the sick role is assumed by persons afflicted by both somatic and mental diseases. In each case, the role is of broad social significance. But in the case of mental illness, the sick role not only excuses one from responsibility, it relieves one of responsibility. At least, this is the case with serious mental illnesses – one is stripped of the usual prerogatives of members of one's society. In fact, being characterized as seriously mentally ill involves the forfeit of many of the rights and duties of persons.

As a result, a philosophical analysis of mental health must also consider the legal and political implications of concepts of mental illness. Chief among these are issues bearing on modes of partial rather than total loss of competency due to mental illness, including different ways of structuring social responses to mental illness. These issues turn on philosophical questions concerning responsibility, accountability, and moral agency. To give an adequate account of mental incompetence, one must determine whether mental competence is either totally possessed or lost, or if it is lost by degrees, one must justify and specify appropriate partial recognitions of responsibility. Further, any theory of partial responsibility has important implications for issues in both civil and criminal law. In addition, such theories guide distinctions between those social interventions which we term 'punishments for crimes' and those we term 'involuntary treatments of the mentally ill.' These issues have been addressed in a series of conferences which were held in the Fall of 1976 at Rice University and the University of Texas Medical Branch on the topic, 'Mental Illness: Law and Public Policy'. The papers from these conferences will appear in the next volume in this series. Here it is enough to indicate the pressing need for a better understanding of the ways in which we should want categories such as 'mentally healthy' and 'mentally ill' to function. These are decisions rich with presuppositions and pregnant with consequences. Part of what philosophy can offer is a greater clarity concerning these presuppositions, and a critical appraisal of the values inherent in a particular view of mental health or illness.

H. T. Engelhardt, Jr. and S. F. Spicker (eds.), Mental Health: Philosophical Perspectives, 295.

NOTES ON CONTRIBUTORS

Baruch A. Brody, Ph.D., is Associate Professor and Chairman, Department of Philosophy, and Director of the Legal Studies Program, Rice University, Houston, Texas.

Chester R. Burns, M.D., Ph.D., is James Wade Rockwell Associate Professor of the History of Medicine and Associate Director of the Institute for the Medical Humanities, The University of Texas Medical Branch, Galveston, Texas.

Daniel L. Creson, M.D., Ph.D., is Associate Professor and Chief of the Division of Community and Social Psychiatry, The University of Texas Medical Branch, Galveston, Texas.

Corinna Delkeskamp, Ph.D., is Adjunct Assistant Professor of Philosophy, Pennsylvania State University, University Park, Pennsylvania.

Alan Donagan, M.A., B.Phil. (Oxon.), is Professor of Philosophy, The University of Chicago, Chicago, Illinois.

H. Tristram Engelhardt, Jr., Ph.D., M.D., was at the time of the conference, Associate Professor, Institute for the Medical Humanities and the Department of Preventive Medicine and Community Health, The University of Texas Medical Branch, Galveston, Texas. He is currently Rosemary Kennedy Professor of the Philosophy of Medicine, Kennedy Institute, Center for Bioethics, Georgetown University, Washington, D.C.

Edmund L. Erde, Ph.D., is Assistant Professor, Institute for the Medical Humanities and the Department of Preventive Medicine and Community Health, The University of Texas Medical Branch, Galveston, Texas.

Horacio Fabrega, Jr., M.D., is Professor of Psychiatry, University of Pittsburgh, Pittsburgh, Pennsylvania.

Leonard Feldstein, M.D., Ph.D., is Professor of Philosophy, Fordham University, New York, New York.

James A. Knight, M.D., is Dean, College of Medicine, Texas A & M University, College Station, Texas.

Karen Lebacqz, Ph.D., is Assistant Professor of Christian Ethics, Pacific School of Religion, Berkeley, California, and Consultant in Bioethics, Department of Health, Sacramento, California.

Ruth C. Macklin, Ph.D., is Associate for Behavioral Studies at the Institute of Society, Ethics and the Life Sciences, and Associate Professor of Philosophy at Case Western Reserve University, Cleveland, Ohio.

Robert C. Neville, Ph.D., is Professor and Chairman of Religious Studies and Professor of Philosophy, State University of New York at Stony Brook, Stony Brook, New York.

Stuart F. Spicker, Ph.D., is Associate Professor of Philosophy and Community Medicine, The University of Connecticut Health Center, Farmington, Connecticut.

Thomas S. Szasz, M.D., is Professor of Psychiatry, State University of New York at Syracuse, Syracuse, New York.

Irving Thalberg, Ph.D., is Professor of Philosophy, University of Illinois at Chicago Circle, Chicago, Illinois.

Stephen Toulmin, Ph.D., is Professor of Philosophy and Professor on the Committee on Social Thought, The University of Chicago, Chicago, Illinois.

Bernard Towers, M.B., Ch.B., is Professor of Pediatrics and Anatomy, Center for Health Sciences, University of California at Los Angeles, Los Angeles, California.

J. H. van den Berg, M.D., Ph.D., is Professor of Phenomenology and Conflict-psychology, State University of Leiden, Leiden, Holland.

Caroline Whitbeck, Ph.D., is Assistant Professor of Philosophy, State University of New York at Albany, Albany, New York.

Richard M. Zaner, Ph.D., is Easterwood Professor of Philosophy and Chairman, Department of Philosophy, Southern Methodist University, Dallas, Texas.

INDEX